T0271853

Non-destructive Diagnostic of High Voltage Electrical Systems: Theoretical Analysis and Practical Applications

RIVER PUBLISHERS SERIES IN ENERGY ENGINEERING AND SYSTEMS

Series Editors

BOBBY RAUF
Semtrain, LLC,
USA

The "River Publishers Series in Energy Engineering and Systems" is a series of comprehensive academic and professional books focussing on the theory and applications behind various energy-related technologies and control systems. The series features handbooks for related technology, as well as manuals on the fundamentals and theoretical aspects of energy engineering projects.

The main aim of this series is to be a reference for academics, researchers, managers, engineers, and other professionals in related matters with energy engineering and control systems.

Topics covered in the series include, but are not limited to:

- Energy engineering;
- Machinery;
- Testing;
- HVAC;
- Electricity and electronics;
- Water systems;
- Industrial control systems;
- Web based systems;
- Automation;
- Lighting controls.

For a list of other books in this series, visit www.riverpublishers.com

Non-destructive Diagnostic of High Voltage Electrical Systems: Theoretical Analysis and Practical Applications

Josef Vedral

Czech Technical University in Prague, Czech Republic

NEW YORK AND LONDON

Published 2023 by River Publishers

River Publishers

Alsbjergvej 10, 9260 Gistrup, Denmark

www.riverpublishers.com

Distributed exclusively by Routledge

605 Third Avenue, New York, NY 10017, USA

4 Park Square, Milton Park, Abingdon, Oxon OX14 4RN

Non-destructive Diagnostic of High Voltage Electrical Systems: Theoretical Analysis and Practical Applications / Josef Vedral.

Routledge is an imprint of the Taylor & Francis Group, an informa business

ISBN 978-87-7022-802-2 (print)

ISBN 978-10-0088-138-7 (online)

ISBN 978-1-003-39419-8 (ebook master)

While every effort is made to provide dependable information, the publisher, authors, and editors cannot be held responsible for any errors or omissions.

Contents

Preface

The book describes the basic aspects of disassembly-free diagnostics of high voltage electrical machines.

The introduction is a general view of the diagnosis of electric machines including off-line and online diagnosis and prognosis of the condition of diagnosed devices. The following chapter is devoted to the disassembly diagnostic of mechanical parts of electrical machines. The properties and types of insulation systems are described, including the methods of electrical stress and degradation caused by their mechanical, temperature and electrical aging. The following chapter describes the measurement methods to measure the properties of these insulation systems. Separate chapters are devoted to the formation and behavior of partial discharges in insulation systems of electrical equipment and to the description of the methods used to measure them. The following chapters describe the voltage testing of electrical machines and diagnostic methods of power transformers, rotary electrical machines, high voltage power cables and insulators including voltage conductors and power shut-offs

The book was created on the basis of the authors' long time cooperation with the company ORGREZ, a.s. within the solution of TAČR (Technology Agenture of Czech Republic) project TA02010311 *"Intelligent measuring diagnostic system for estimating the operating condition of high-voltage electric machines"* and the related project TH02020288 *"Compact diagnostic system for monitoring the condition of high-voltage electric machines using DC and low-frequency alternating test voltage"*.

In connection with the publication of the book, the authors would like to thank in particular Ing. Jiří Brázdil, Ph.D., MBA, director of the division of electrical laboratories of ORGREZ company, Ing. Lubomír Markus, MBA, head of the diagnostics and high voltage testing department from the same workplace, František Antfeist, former head of the technology and diagnostics department of ČEZ (Czech Energy Company), for consultations and inspiring comments on the content of the book. Furthermore, the authors thank doc. Ing. Radek Procházka, Ph.D. from the Department of Electrical Power Engineering, Faculty of Electrical Engineering, Czech Technical University in Prague for consultations and assistance in verifying the properties of devices for measuring partial discharges, Ing. Radek Sedláček, Ph.D., Ing. Ján Tomlain and Ing. Ondřej Tereň from the Department of Measurement, Faculty of Electrical Engineering, Czech Technical University in Prague for long-term cooperation in the development of diagnostic devices within the above-mentioned TACR projects.

I would like to thank particularly may wife Helena and may all family for moral and factual help in writing and editing of books.

Prague, December 2022
Josef Vedral

List of Figures

List of Tables

List of Abbreviations

ADC	Analog-to-digital converter
CCD	Charge coupled device
CCD	Charge coupled devices
CSM	Chlorosulfonated polyethylene
CT	Crest factor
DA	Differential amplifier
DAR	Dielectric absorption ratio
DD	Dielectric discharge
DGA	Dissolved gas analysis
DMSO	Dimethyl sulfoid
DUT	Devices under test
EP	Epoxies
EPM	Ethylene-propylene elastomers
FDS	frequency domain spectroscopy
FID	Flame ionization detector
FPGA	Field programmable gate array
FRA	Frequency response analysis
GC	Gas chromatography
GST	Grounded specimen test
GSTQ	Grounded specimen test quarding
HV	High voltage
HVDC	High voltage direct current
IEC	International Electrotechnical Commission
IFRA	Impulse frequency response analyzing
IPC	Integrated circuit piezoelectric
ISO	International Organization for Standardization
LP	Low-pass
MEMS	Micro electro mechanical systems
MV	Mega Volt
MVA	Mega volt amp
NTC	Negative temperature coefficient
PCA	Polycyclic aromatics

PCB	Polychlorinated biphenyl
PDM	Partial discharge meter
PE	Polyethylene
PE	Polyethylene
PI	Polarization index
PLM	Polyamides
PP	Polypropylene
PTC	Positive temperature coefficient
PTFE	Polytetrafluoroethylene
PUR	Polyurethane
PVC	Polyvinyl chloride
RMS	Root mean square
RPG	Rectangular pulse generator
SD	Schottky diode
SFRA	Sweep frequency response analyzing
SG	Signal ground
TCD	Temperature compensated diode
UHF	Ultra-high frequency
UHV	Ultra-high voltage
UST	Unground specimen test
VHV	Very high voltage
VLF	Very low frequency
XLPE	Cross-linked polyethylene
XRF	X-ray fluorescence

1

Introduction

Non-disassembly diagnostics of mechanical and electrical parts of HV rotating and non-rotating electrical machines (generators, motors, transformers, etc.) date back to the 1930s in connection with the mass production and use of electricity[1].

The technical condition of the *mechanical parts* of rotating HV electrical machines is determined mainly by their *vibrations*, which are caused by the imbalance of the rotating parts of the machines and the condition of their bearings. In connection with the geometric inaccuracies of their stators and rotors, *shaft voltages* and *currents* arise, which cause wear of machine bearings (electrical erosion). Torsional oscillations of their shafts can also occur in large MV electric rotary machines with long shafts, e.g., steam generators. Mechanical, magnetic, or optical methods are used to measure them.

The *technical condition* of HV electrical parts of rotating and non-rotating electrical machines is further determined mainly by the state of their insulation, i.e., *leakage current, insulation resistance, capacity, loss factor, dielectric absorption of the dielectric*, and the *occurrence of partial discharges*. Special meters of insulation, capacity, loss factor, and dielectric absorption are used to measure these parameters.

A very widespread diagnostic method for determining the insulation status of large HV electrical rotating and non-rotating machines is the measurement and detection of partial discharges in their insulation systems.

Galvanic methods are used for contact measurement of partial discharges, in which the HV excitation voltage is reduced by an HV capacitive divider and voltage pulses, the area of which corresponds to the magnitudes

---

[1] In terms of voltage levels, we distinguish small voltages SV < 50 V, low voltage LV 50 V to 1 kV, high voltage HV 1–50 kV, very high voltage VHV 50–400 kV, and ultra-high voltage UHV 400–800 kV.

1

of the partial discharges, which are integrated in the given frequency band. *Through-current current transformers* are used for *non-contact measurement* of partial discharges, which are threaded on conductors through which excitation currents of diagnosed machines flow and which also contain current pulses, whose charges correspond to the measured partial discharges. *Inductive, capacitive*, and *high frequency probes* are used to detect partial discharges, which are placed either directly in the insulation systems of electrical machines or in their immediate vicinity. Non-electrical methods of partial discharge detection include ultrasonic, optical, and chemical methods.

In addition to measuring capacity changes and loss factor, chemical analysis of their cooling oils is also used to diagnose the insulation status of HV and VHV transformers. Methods using shock waves and frequency analysis of winding transmission impedance are determined to identify faults and between short-circuit shorts of transformer windings. To identify faults and the place of their occurrence, suitable methods for measuring current leakage and refractometric methods for measuring the reflection of impulse excitation signals are used in HV cables.

The task of diagnostics of large HV electrical machines is not only to find out their technical condition but also to estimate their time of trouble-free operation. We recognize initial, operational, one-time, periodic, and permanent diagnostics of mechanical and electrical parts of machines. We also distinguish between assembly and non-disassembly diagnostics and offline and online diagnostics, of which non-disassembly online diagnostics enable permanent and efficient determination of parameters of strategic HV electrical machines and equipment, whose trouble-free operation is crucial for power plants, distribution and traction transformers, oil pipeline pumps, gas pipeline compressor stations, etc.

In addition to the diagnosis itself, each successful diagnostic also works with a prognosis, i.e., determination of the future development of the technical condition of the object and genesis, i.e., sufficient knowledge of the "history" of the monitored object. Therefore, sophisticated methods are used to process diagnostic signals, using elements of artificial intelligence using neural networks. Comparative databases of relevant programs, created based on long-term experience gained in the diagnosis of HV electrical machines, are highly valued knowledge of companies engaged in the development, production, and use of these diagnostic systems in practice. The results obtained in this way make it possible to more accurately estimate not only the development trend of machine failures but also the date of their eventual accident.

Figure 1.1 shows a typical time course of the frequency of failures of large energy machines and equipment.

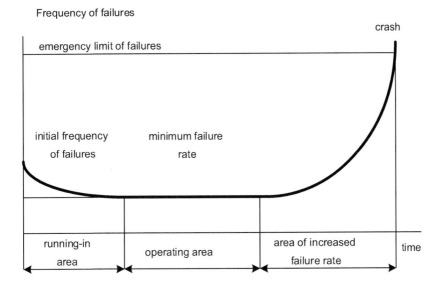

Frequency of failures

crash

emergency limit of failures

initial frequency   minimum failure
of failures   rate

running-in   area of increased   time
area   operating area   failure rate

Figure 1.1 Time dependence of the frequency of failures of electrical machines.

The course of the frequency of failures of these machines or the device usually has three time intervals.

In the first time interval, the so-called run-in areas and the frequency of their failures decrease. Inaccurate assembly of their mechanical parts causes improper assembly of bearings and polarization of dielectrics of their insulation systems these faults, for example. This time can take several hours to several tens of hours of operation.

In the second time interval, the so-called operating area, when the main causes of the *initial failures* are already eliminated, the frequency of machine failures is kept to a minimum. At the end of this time interval, this frequency begins to increase due to the limited service life of the mechanical and electrical parts of the machines. At that time, it is advisable to carry out partial or complete repairs to extend the life of the machines. If repairs are not carried out in time, the frequency of failures will reach the emergency limit, leading to a machine accident. Based on this trend in the frequency of failures, it is then possible to estimate the effective life of machines or the date of their accident.

For example, large rotating HV electrical machines have a design life of 20 years. With proper maintenance and diagnostics, this service life can be doubled. A standard high-voltage MV transformer has a design life of 30 years. With regular maintenance and diagnostics, this service life can be extended up to 50 years. HV distribution cables have a design life of up to 50 years; their actual life can be even higher depending on the type and method of their load.

Monitoring the intensity of failures is essential to plan machine repairs and eliminate their accidents. Because unplanned machine downtime is very costly, these machines must be diagnosed regularly and the results obtained are used to identify the type and extent of faults. This can effectively prevent significant economic losses during downtime. Similar financial losses will be caused by unplanned outages of main HV distribution transformers or HV ground cables.

2

General View on Diagnostics of Electrical Machines

The increasing demands on the operability, reliability, and quality of all electrical equipment and the reduction of their production and operating costs require the use of sophisticated methods of their diagnostics. These requirements are met by a discipline called *electro-technological diagnostics*, which is closely related to the diagnostics of mechanical and electrical properties of HV electrical machines.

The evaluation of the service life and reliability of HV electrical equipment is conditioned by the knowledge of determining the causes and effects of their failures. The resulting faults are recorded and stored in *knowledge databases*. Based on their evaluation, the diagnostic system determines the places of failure, the methods of their repair, and the possibilities of their prevention in the event of subsequent fault conditions. Mathematical and physical models are used to diagnose the condition of MV electrical equipment, determining the residual life of HV equipment.

2.1 Diagnosed Object

We consider a diagnostic object to be a system that consists of interconnecting functional elements and subsystems. These can be considered as a set of n elements, which are affected by a set of external influences and stimuli, to which the system responds by a set of its reactions (output signals). During their life, the diagnosed objects are mostly in a *fault-free state*, when they are able to perform all functions and the output signals are within tolerance.

The opposite is a fault condition when the output signals are out of tolerance and the object is unable to perform its function. If the object has the output signals of the diagnostic device outside the tolerance limits but is able to fulfill its function, we speak of a serviceable condition.

We consider the conditions for the operation of the equipment of the diagnosed object to be mostly normal. These are conditions that respect the requirements of the manufacturer. If a failure occurs in such a situation, it means that it is caused by poor construction or design of the object. If the object is operated outside the conditions specified by the manufacturer, these are abnormal conditions and failures may be caused by external influences.

Diagnosed objects are divided into objects with accessible and inaccessible structure in terms of their structure. *Objects with an inaccessible structure* can monitor the behavior of objects only through input and output signals, while they do not allow monitoring processes in their internal structure. *Objects with an accessible structure* also allow you to monitor the behavior of an object within their structures.

The carrier of information on the technical condition of the diagnosed object is a *diagnostic quantity* that can transform parameters that are not directly measurable into a measurable state. The diagnosed object is determined by its states, which are determined and recorded by *diagnostic signals*. A function block that converts a set of function states to a set of function signals is called a diagnostic system. With its help, we are able to measure diagnostic signals and obtain data on the status of the diagnosed object. The following resources must be available for the diagnostic system to function properly.

The *necessary instrumentation consists* of a set of meters with suitable converters needed to convert a set of object states to a set of signals that we measure and record. The sensor system should be part of the diagnosed object, as it is necessary to perform diagnostics during the construction of the electrical equipment.

Model of the diagnosed object for simulation of all operating conditions that may occur during the production and operation of the diagnosed object. During the preparation of the model, the characteristics of the diagnosed object, its characteristic states, and mathematical expression of its parameters are collected. Such models can be *physical* (usually direct copies of objects or parts of the object on which diagnostics are performed) or *mathematical*. The basis of the model is a block diagram, which clearly shows all input and output signals of the function blocks of the object.

The approach to solving the diagnosed object can be either phenomenological or structural. The *phenomenological model* means that during diagnostics, we are only interested in the object's reactions to the input diagnostic signals. In the *structural model*, we examine the events in the structure of the diagnosed object. It follows that in a structural approach, we obtain much more information about an object with less variance of values. However, such an approach is much more complex and expensive for instrumentation and its operation.

The *decision on the type of test* can be destructive or non-destructive. The *destructive test* has an excellent informative value, but it is demanding on the consumption of experimental material, which is devalued by the test. In contrast, non-destructive tests can be repeated as the test material remains almost undamaged. *Non-destructive testing* can therefore be used in online diagnostics.

Determining the procedure and knowledge potential consists in determining one's own diagnostic operations, optimizing them and determining the individual steps of the diagnosis, and it is equally important to identify employees who have the necessary knowledge and experience.

2.2 Online and Offline Diagnostics

In terms of the *choice of diagnostic methods*, we distinguish between online and offline diagnostic methods. *Online diagnostics* are performed during full operation of the device and the functionality of the diagnosed system is almost independent of this mode. *Offline diagnostics* are performed when the device is shut down. During this diagnosis, in most cases, the device is disassembled or some intervention is performed on the diagnosed object.

For a general description of both methods, the *transfer function* of the object is compiled and, initially, its simple model is used. The *transfer function* expresses the dependence of the output functions $Y = (y_1, y_2, \ldots, y_n)$, realized by the diagnosed object on the set of its input variables $U = (u_1, u_2, \ldots, u_n)$ and internal states $X = (x_1, x_2, \ldots, x_n)$ at time t.

Applies to *fault-free condition*

$$Y = \Psi(U, X_p, t),\qquad(2.1)$$

where X_p is the initial state value of the diagnosed object.

It applies to the *i*th fault condition

$$Y_i = \Psi^i(U, X_p, t).\qquad(2.2)$$

Checks $K = \{k_1, k_2, \ldots, k_n\}$ are performed on the diagnosed object, which are obtained on the basis of partial values of test signals aj. The response of the object to the partial control k_j is given by the set of measuring points $\{\gamma\}_j$. The result of the partial check is the technical state of the object R_j^i. The result of the partial check, which is a function of the influence j, can generally be written as a sequence $\{\gamma\}_j$ of j-dimensional phasors

$$R_j^i = \Psi^i\left(\alpha_j, \{y\}_j\right), \tag{2.3}$$

where $R_j = \Psi(k_j)$ holds for the *fault-free state* of the object, and $R_j^i = \Psi^{ij}(k_j)$, for the object in the ith fault state.

2.2.1 Offline diagnostics

The block diagram of offline diagnostics is shown in Figure 2.1.

In the *control block*, the diagnosis algorithm is stored. According to his commands, the signal source generates signals of partial controls α_j in a sequence that corresponds to the sequence of the diagnosis algorithm.

In the *control block*, the diagnosis algorithm is stored. According to his commands, the signal source generates signals of partial controls α_j in a sequence that corresponds to the sequence of the diagnosis algorithm. These

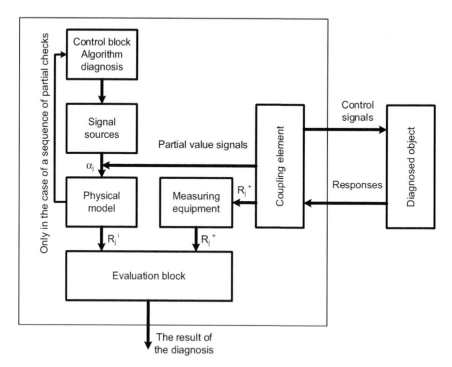

Figure 2.1   Block diagram of offline diagnostics.

signals are fed into the physical model and through the coupler to the diagnosed object. The response of these signals in the physics model $\{R_j^i\}$ is compared in the evaluation block with the response of the signals from the diagnosed object $\{R_j^+\}$. Based on the comparison of these signals, a diagnostic result can be determined.

2.2.2 Online diagnostics

Functional online diagnostics takes place during the full operation of the diagnosed object. The block diagram of the online diagnostics is shown in Figure 2.2.

At the input of the *diagnosed object* come working signals. On their basis, the diagnostic object generates control signals y_j and responses $\{R_j^+\}$. These signals are fed through the coupler and measuring device to the evaluation block, where they are compared with the $\{R_j^i\}$, signals that are generated by the physical model of the diagnosed object. Based on the comparison of the signals $\{R_j^+\}$ and $\{R_j^i\}$, the result of the diagnosis of the object is determined.

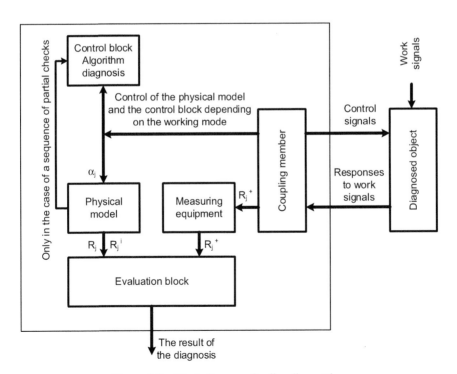

Figure 2.2 Block diagram of online diagnostics.

2.3 Forecast of the Condition of the Diagnosed Device

The task of establishing a forecast is to determine the development of the state of the diagnosed device. If the current state of the diagnosed object and the previous development of its state are known, we can predict the future behavior of the object, i.e., to predict the development of its parameters.

If we consider the disorder condition as a phenomenon that occurred by the gradual deterioration of its parameters, it is a *gradual disorder*. Based on the knowledge of the course of the change in the state of the object, we can prevent another possible malfunction. The task of fault prediction is therefore to determine the deterioration of the condition of the object with the help of available means and thus prevent an unexpected failure.

If the state of the diagnosed object is determined by its parameters $x = (x_1, x_2,\ldots, x_i)$, then its state can be expressed by time dependence

$$x(t_i) = \varphi(t_i). \tag{2.4}$$

Time moments of operation of the diagnosed device $t_1 < t_2 < \cdots t_n$ correspond to device states $x(t_1), x(t_2),\ldots, x(t_n)$.

The most widely used ways to forecast the condition of equipment is the analytical and statistical method.

The analytical method of forecasting determines the dependence of the function $x(t_i)$ on the monitored parameter x_i (Figure 2.3).

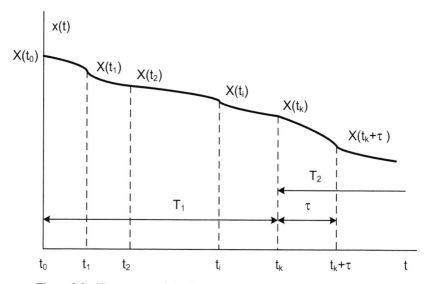

Figure 2.3   Time course of the dependence of the monitored parameter $x(t)$.

The state of the entire system, consisting of the elements of the system, is expressed by the parameters defining the states of these subsystems. In the time range T_1, the monitored function takes the values $x(t_0)$, $x(t_1)$, ..., $x(t_k)$, where $t_0 < t_1 < \cdots < t_k$.

Based on the known values of $x(t_i)$, it is necessary to predict the values of the function $x(t_k + 1)$, $x(t_k + 2)$,..., $x(t_k + m)$ for moments $t_k + 1$, $t_k + 2$,..., $t_k + m$, where $t_{k+1} < t_{k+2} < \cdots < t_{k+m}$.

According to this method of forecasting, predicting the behavior of a diagnosed object is a prediction of a change in the x_j parameter, expressing the state of the device, which at some point reaches the critical value of x_{ikr}. The prediction is derived from the function directive

$$\frac{dx_i(t)}{dt} = \frac{x_i(t_k) - x_i(t_{k-1})}{t_k} - t_{k-1} \tag{2.5}$$

and equation

$$x_i(t_k + \tau) = x_i(t_k) + \frac{dx_i(t)}{dt}, \tag{2.6}$$

where t is the following time interval.

To achieve the critical value of x_{ikr}, we assume the time

$$t_k + \tau = \frac{x_i - x_i(t_k)}{\dfrac{dx_i(t)}{dt}}. \tag{2.7}$$

The *statistical method of evaluating the forecast* is based on the assumption that on the basis of the known values of $x(t_i)$, it is possible to determine the probability that the function $x(t)$ will not exceed the limit of the permissible values

$$P(x) : \left[\left| x(t_{k+m}) - x_{nom}(t) \right| \preceq \xi_{perm} \right] \tag{2.8}$$

where t_i are the time moments from the region T_1 ($i = 0, 1, 2,...,k$), $x(t_{k+m})$ are the values of the parameter of interest at the time moments t_{k+m} of the region T_2, $x_{nom}(t)$ is the nominal value of the parameter, and ξ_{perm} is the *permissible deviation* of the function $x(t)$ in the time section T_2.

This approach to system state forecasting assumes that during the assignment of values to system states, it is possible to consider not only the statistical parameters of states but also their transition states. The probability that the function $x(t)$ does not exceed the limit of the permissible values can then be expressed by the following equation:

$$P(x) : \left[x_{\text{nom}} < x(t_{k+m} + \xi_{\text{dov}} \right] = \int_{x_{\text{nom}}}^{x_{\text{nom}}+\xi} f\left[x(t) \right] dt \tag{2.9}$$

The solution to predict the behavior of the diagnosed object consists of the following two stages.

Collection of a sufficient number of data, i.e., monitoring the genesis of the behavior of the diagnosed object, their processing, and evaluation. This gives an idea of the tendency of the change of the monitored parameter and thus the determination of a more precise pattern of change of the monitored function $x(t)$, including the choice of a suitable prediction step.

The actual processing of the forecast by the appropriate procedure, in which it is possible to select the methods of calculating the observed trend according to the nature of the case and the required accuracy of the forecast determination.

References

[1] Gertler, J., *Fault detection and diagnostics in engineering systems.* Dekker, 1998. ISBN: 0-8247-9427-3
[2] Patton, R., Frank, P. Clark, R. *Fault diagnostic in dynamic systems.* Springer Verlag 2006. ISBN: 978-3-540-2412-6
[3] Dursun, D. *Predictive analysis.* KNIME 2019. ISBN: 978-0136738510
[4] Masmadijan, D. Farnod, R. *The art of modeling in science and engineering with mathematica.* CRC Press, 2006. ISBN-13: 978-1584884606
[5] Tzafestas, S. *Expert systems in engineering applications.* Springer Verlag.1993. ISBN: 97-8364284-0500
[6] Giarranto, J.c., Riley, G. *Expert systems. Principles and programming.* PWS Publishing, 1998. ISBN: 97-80534-9538

3

Diagnostics of Mechanical Parts of HV Machines

Typical areas of condition monitoring of large HV electrical machines are condition monitoring of their mechanical parts. This area of diagnostics includes the measurement of machine vibrations, the torsional oscillation of their shafts, the measurement of bearing temperatures, and temperature images of electrical machine parts.

3.1   Measuring Machine Vibration

Vibration measurement and its effect on the technical condition of machines are dealt with by a field called *vibrodiagnostics*. The cause of vibration of rotating electric machines, e.g., turbo-generators and engines, is the dynamic imbalance of their rotors, the condition of bearings, gearboxes, and other moving parts of their parts. The cause of vibrations of non-rotating electric machines, e.g., transformers and converters, is their excitation by alternating electric current.

3.1.1 Types of vibrations and vibrodiagnostic quantities

The *vibrations of machines* and their parts are manifested by their oscillation, which can be periodic, non-periodic, or random. *Periodic oscillations* at one frequency are referred to as *harmonic oscillations*. *Non-harmonic oscillations* are formed by the superposition of several harmonic components of oscillation and can be either periodic or non-periodic.

   Periodic non-harmonic oscillations arise from the superposition of harmonic oscillations, and their frequencies are in the ratio of rational numbers. If the frequencies of the components are in the ratio of irrational numbers, then we refer to this oscillation as *non-periodic non-harmonic oscillations*.

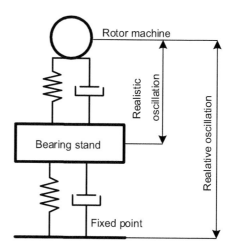

Figure 3.1   Absolute and relative oscillations.

According to the reference point to which we refer to oscillation, we distinguish absolute and relative oscillations (Figure 3.1).

Absolute oscillation refers to the gravitational field of the globe (fixed point), which can be, for example, the frame of a machine connected to the ground.

Relative oscillation refers its quantities to the selected reference point of the machine, e.g., a bearing stand that is not firmly connected to the ground and can also be in oscillating motion relative to the fixed point. To express the properties and effect of vibrations on machines, instantaneous deflection, *amplitude, oscillation*, and *oscillation phase* are used.

In addition, the *speed of oscillation*, its *acceleration*, and the time change of acceleration, referred to as *shock*, are assessed. The integral quantities of oscillation are its *mean* and *RMS values*. The square of the effective value, referred to as the *cardinality* of the oscillation, expresses its performance. The derived quantity of oscillation is the crest factor of oscillation, given by the ratio of the maximum value to the mean value of harmonic oscillation.

Table 3.1 provides an overview of vibrodiagnostic quantities used to express oscillation properties.

Figure 3.2 shows the time course of the harmonic oscillation deflection with its typical quantities.

For the harmonic signal, the oscillation of the deflection $X_r j$ is equal to twice the amplitude of the deflection X_m.

For mean and effective values of deflections

$$X_s = \frac{2}{\pi} X_m = 0.637 \cdot X_m, \qquad X_{ef} = \frac{X_m}{\sqrt{2}} = 0.707 \cdot X_m. \qquad (3.1a, b)$$

Table 3.1 Vibrodiagnostic quantities.

Brand	Unit	Title	Definition
x	m	Immediate deflection	Distance of the oscillating point from the reference point
X_m	m	Amplitude of oscillation	Maximum value of harmonic oscillation deflection
X_r	m	Amplitude	Maximum difference of maximum values in a given time interval
φ	rad	Phase	Delay from $t = 0$
ω	rad·s^{-1}	Angular frequency	Phase change in time $\omega = d\varphi/dt$
T	S	Period	Reciprocal frequency value $T = 1/f$
f	Hz	Frequency	Reciprocal value of the period $f = 1/T$
v	ms^{-1}	Speed	Time change of deflection $v = ds/dt$
a	ms^{-2}	Acceleration	Time change of speed $a = dv/dt$
b	ms^{-3}	Shock	Acceleration time change $b = da/dt$
X_s	m	Mean value of deflection	$X_s = \dfrac{1}{T}\displaystyle\int_0^T \lvert x(t) \rvert\, dt$
X_{ef}	m	Effective deflection value	$X_{ef} = \sqrt{\dfrac{1}{T}\cdot x^2(t)\, dt}$
CT	–	Crest factor	$CT = \dfrac{X_m}{X_{ef}}$

3.1.2 Vibrometer sensors

The principle of operation of classical *seismic oscillation sensors* is the movement of a body with a mass m in relation to a reference body with a mass of M (Figure 3.3).

The motion of a body of mass m is defined by the differential equation of motion

$$m \cdot \frac{d^2 y(t)}{dt^2} + b \cdot \frac{dy}{dt} + k \cdot y = F_b = M \cdot \frac{d^2 x(t)}{dt^2}, \qquad (3.2)$$

where y is the deflection of a body of mass m, dy/dt is its speed of motion, d^2y/dt^2 is its *acceleration* with respect to a reference object of mass M, b is the *damping coefficient*, k is the *stiffness of the springing* element, and F_b is

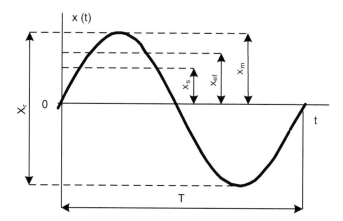

Figure 3.2　Time course of harmonic oscillation deflection.

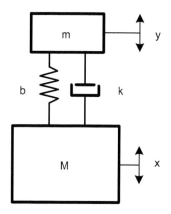

Figure 3.3　Principle of operation of the vibration sensor.

the force acting on a body of mass m. By selecting the parameters m, b, and k, it is then possible to create sensors of deflection, speed, and acceleration.

　　Vibrometers are characterized by a large mass m, comparable to the mass M of the reference object, small damping b, and low stiffness to the springing element. For these sensors, the equation of motion approximately applies

$$y(t) \approx x(t). \tag{3.3}$$

Therefore, the deflection of a body with a mass of m is approximately equal to the deflection of the reference body for these sensors

$$m \cdot \frac{d^2 y(t)}{dt^2} \approx M \cdot \frac{d^2 x(t)}{dt^2}. \tag{3.4}$$

Velocimeters are characterized by a significantly small mass m, a large damping b, and a large stiffness to the springing element k. For these sensors, the equation of motion approximately applies

$$b \cdot \frac{dy}{dt} \approx M \cdot \frac{d^2 x(t)}{dt^2}, \tag{3.5}$$

and, therefore, the deflection of a body with a mass of m is proportional to the speed of movement of the reference body for these sensors

$$y \approx \frac{M}{b} \cdot \frac{dx(t)}{dt}. \tag{3.6}$$

Accelerometers have considerable spring stiffness and a small mass m. For these sensors, the equation of motion approximately applies

$$ky \approx M \cdot \frac{d^2 x(t)}{dt^2}, \tag{3.7}$$

and, therefore, the deflection of the body with mass m is proportional to the acceleration of the reference body for these sensors

$$y \approx \frac{M}{k} \cdot \frac{d^2 x(t)}{dt^2}. \tag{3.8}$$

3.1.2.1 Deflection sensors

Inductance sensors are most often used to detect deflections. These sensors take advantage of the dependence of the change in the inductance of the coil with an open ferromagnetic core on its distance from the electrically conductive material whose movement we are detecting. Eddy currents, caused by the high-frequency magnetic field in this material, create a secondary magnetic field in it, the intensity of which counteracts the intensity of the magnetic field of the excitation coil, causing a reduction in the impedance of the telecoil.

If the excitation coil is, at the same time, a coil of the oscillating circuit, *amplitude modulation* of the oscillating voltage occurs during the oscillating movement of the measured object. By demodulation of this course, we obtain a voltage that corresponds to the time course of the deflection of the measured object.

In order to eliminate external interference, the oscillating coil including the *amplitude demodulator* is placed in an electrostatic (metal) shielding cover. The frequency range of these sensors is 10 kHz, with zero frequency corresponding to the static measurement of the movement of the object (Figure 3.4).

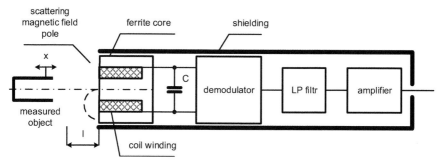

Figure 3.4 Inductive vibration sensor.

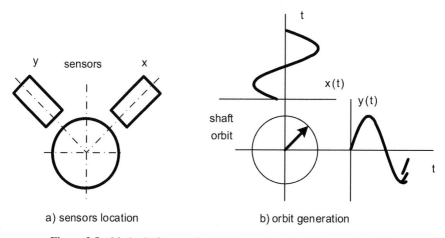

a) sensors location   b) orbit generation

Figure 3.5 Method of measuring the kinematic path of the rotating shaft.

If we place two of these sensors near the shaft of the rotating machine so that their axes are 90°, it is possible to evaluate the *kinematic path* of the machine shaft, the so-called *orbit* (Figure 3.5).

From the course of the kinematic path of the shaft, it is possible to determine the instantaneous values of deflections x and y on each other perpendicular directions and the instantaneous kinematic deflection of the shaft, given by the vector sum of the two deflections.

3.1.2.2 Speed sensors

A typical *absolute speed sensor* is a sensor with a moving coil in which a permanent magnet moves (Figure 3.6).

The coil is suspended on a spring or on a flexible membrane. To dampen the movement of a permanent magnet, either a damping fluid in which the

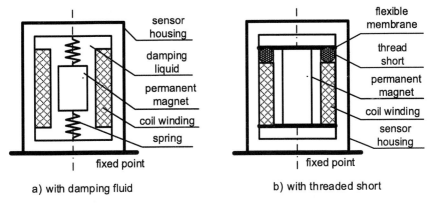

a) with damping fluid   b) with threaded short

Figure 3.6 Electrodynamic speed sensors.

magnet moves or a short thread that is placed in the coil is used. In both cases, the movement of the magnet in the coil induces a voltage in it

$$u(t) = B \cdot l \cdot v, \tag{3.9}$$

where B is the induction of the magnetic field of the permanent magnet, l is the length of the conductor in this field, and v is the speed of movement of the coil.

This electrodynamic sensor is inherently an *absolute sensor*, but due to its mechanical–electrical way of converting the signal, the voltage on the coil is directly proportional to the speed of oscillation of its housing. These sensors are able to measure the oscillation rate in the range of several mm s^{-1} to several m s^{-1} in the frequency range of several Hz to several kHz. By integrating the voltage of these sensors, information about their movement can be obtained.

3.1.2.3 Acceleration sensors

To measure acceleration, piezoelectric, piezoresistive, or capacitive accelerometers are most often used.

Piezoelectric accelerometers consist of a seismic body, which is attached to the sensor housing by one end through the springing and damping layer and the other end is attached to the piezoelectric sensor (Figure 3.7).

If a seismic body with a mass m of acceleration a is applied to a *seismic body*, then the resulting force F acting on the *piezoelectric transducer* generates a charge between its electrodes.

$$q(t) = K_p . F(t) = K_p . m . a(t), \tag{3.10}$$

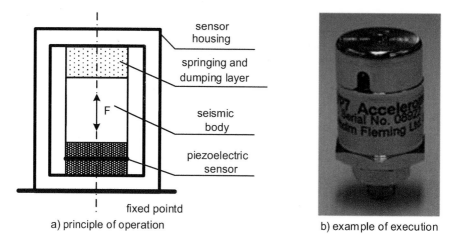

a) principle of operation   b) example of execution

Figure 3.7   Piezoelectric accelerometer.

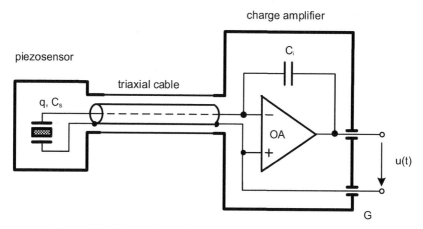

Figure 3.8   Connecting a piezoelectric sensor to a charge amplifier.

where k_p is the sensitivity of the piezo sensor. This charge is converted to voltage by a *charge amplifier* consisting of an operational amplifier with a feedback capacitor C_i (Figure 3.8)

$$u(t) = \frac{q(t)}{C_s} = K_p \cdot \frac{m}{C_s} \cdot a(t). \qquad (3.11)$$

Assuming that the *operational amplifier* OA used has a sufficiently large differential gain, of the order of 10^5 and 10^6, then its input differential voltage is approximately zero, and, therefore, the influence of leakage resistance and

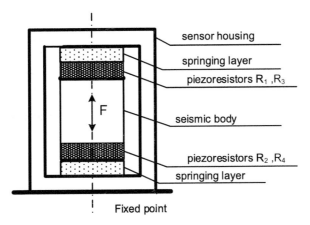

Figure 3.9 Accelerometer with piezoresistors.

capacitance between the inner conductor of the *coaxial cable* and its sheath forming the signal ground of the sensor does not apply. In order to eliminate the effects of external electromagnetic interference, both the sensor and the cable sheath, including the charge amplifier, are shielded and grounded.

To connect the piezoelectric sensor to the charge amplifier, special triaxial cables with minimal *piezoelectric* and *triboelectric effect* are used, which are the cause of the formation of interfering charges during their mechanical stress. Modern piezoelectric accelerometers, referred to as **IPC** (*integrated circuit piezoelectric*), have a charge amplifier located directly in the accelerometer housing and, therefore, do not require their connection with special and expensive cables.

Piezolectric accelerometers allow you to measure acceleration in the range of 1–50 m·s⁻²; their sensitivity reaches several tens to hundreds of mV/ ms⁻² and their frequency range is in the range of several Hz to tens of kHz.

Piezoresistive accelerometers also consist of a seismic body that is attached to the sensor housing via springing and damping layers (Figure 3.9).

Two pairs of *piezoresistors* with identical properties and nominal values are stored between the springing and damping layers, and these pairs are alternately mechanically stressed to tension and pressure by the time change of the measured acceleration. These piezoresistors are connected to the bridge for whose differential voltage (Figure 3.10)

$$u_d(t) = k_p \cdot \frac{dR}{R} \cdot U_n,$$ (3.12)

where k_p is the sensitivity of the piezoresistor, dR/R is the relative change in its resistance, and U_n is the supply voltage of the bridge.

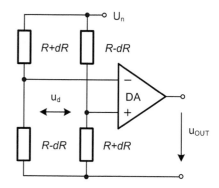

Figure 3.10   Connection of piezoresistors to the bridge.

Since the change in the resistance of piezoresistors is directly proportional to their mechanical stress, the difference in bridge stress is

$$u_d(t) = U_n \cdot k_r \cdot F(t) = U_n \cdot k_r \cdot m \cdot a(t), \qquad (3.13)$$

where U_n is the supply voltage, and k_r is the *resistance sensitivity* of the piezoresistive sensor.

The difference voltage, reaching a maximum of several tens of mV, is amplified by the *differential amplifier* DA to the desired value. In addition, by connecting piezoresistors to the bridge, their temperature dependence is eliminated.

Capacitive accelerometers use *micro electro mechanical systems* (MEMS) micromechanical technology to measure acceleration. This technology uses differential ridge condensates to measure the movement of electrodes to which seismic bodies are attached (Figure 3.11).

The change in the measured acceleration causes changes in the distances of the electrodes $d \pm \Delta d$ of the differential capacitors from their medium electrodes, causing changes in their capacities $C \pm \Delta C$. These capacitance changes are converted by the *charge amplifier* into voltage changes

$$\Delta u = 2 \cdot \frac{\Delta C}{C_i} \cdot U, \qquad (3.14)$$

where C_i is the integration capacity of the charge amplifier.

Capacitive accelerometers are produced in one-, two-, or *three-axis design*, which allows their use even in the vector measurement of vibrometric quantities. Capacitive accelerometers allow you to measure acceleration in the range of several ms^{-2} to 100 ms^{-2}; their sensitivity reaches several tens to hundreds of mV/ms^{-2} and their frequency range is several tens of kHz.

Figure 3.12 shows typical three-axis accelerometers used to measure vibration in industrial practice.

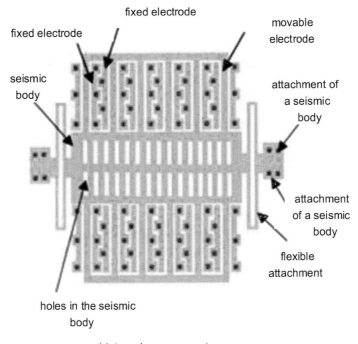

a) internal arrangement

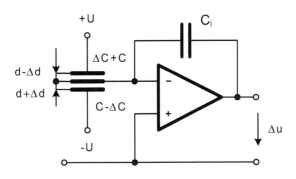

b) differential charge amplifier

Figure 3.11   MEMS accelerometer.

To mount the sensors, either stud screws are most often used, while the attachment surfaces of the sensors are provided with thin *silicone jelly*, or the sensors are glued with a special *cyanoacrylate glue*. Some sensors are also equipped with permanent magnets, which allow them to be easily connected to ferromagnetic materials.

a) accelerometer for gluing   b) accelerometer for mounting

Figure 3.12   Three-axis MEMS accelerometers.

In connection with the attachment of the sensors, it is also necessary to consider the influence of their weight on the determination of vibration quantities of the oscillating parts of the machines. If the mass of the accelerometer m_a is not significantly less than the weight of the measured machine m_s, then for the acceleration of the oscillation of the machine

$$a_s = \frac{m_s + m_a}{m_s} \cdot a_m,$$ (3.15)

where a_m is the measured acceleration by the accelerometer.

3.1.3 Analysis of vibrometric signals

Signals from vibrometric sensors are evaluated either in the time or frequency domain.

3.1.3.1 Analysis of vibrometric signals in the time domain

In the time analysis of vibrometric signals, either the energy of the total oscillation, determined by the square of the effective value of the oscillation rate (the *cardinality of the oscillation*), is evaluated, or the time waveforms of individual vibrometric quantities are evaluated.

Based on the comparison of these quantities, measured on a new or trouble-free machine with the quantities measured on the diagnosed machine, the technical condition of the machine in the frequency range of 10 Hz to 1 kHz is determined on the basis of experimental knowledge.

According to the recommendations of ISO 2372, the condition of the machines is evaluated according to Table 3.2, in which the limits of the permissible magnitude of oscillation are determined according to the performance of the machine.

Table 3.2 Evaluation of the machine according to ISO 2372 recommendations.

Comment	Category A (to 15 kVA)	Category M (15 kVA to 75 kVA)	Category G (> 75 kVA)
Unacceptable oscillation	4.5 to 50 mm s^{-1}	7.1 to 50 mm s^{-1}	11.2 to 50 mm s^{-1}
Oscillations on the admissibility	1.8 to 4.5 mm s^{-1}	2.8 to 7.1 mm s^{-1}	4.5 to 11.2 mm s^{-1}
Permissible oscillation	0.7 to 1.8 mm s^{-1}	1.1 to 2.8 mm s^{-1}	1.8 to 4.5 mm s^{-1}
Small oscillation	0.18 to 0.7 mm s^{-1}	0.18 to 1.1 mm s^{-1}	0.18 to 1.8 mm s^{-1}

To evaluate the state of the machines, in addition to the magnitude of the oscillation, the amplitude, oscillation, and the factor of the oscillation are also assessed. The amplitude of the oscillation determines, to some extent, the overall safety of the machine. Oscillation is suitable for the assessment of structural clearances and mechanical stress of the material.

From the time course of the vibration monitoring and the effective value, the initial damage to the machine can be determined. In this case, the effective value of the oscillation increases later before the increase in the oscillation.

The extreme of the *crest factor* (CT) fluctuation factor then indicates the occurrence and the extent of damage to the machine (Figure 3.13).

To diagnose the *condition of rotating machine bearings*, methods are often used in which the ratio of initial and operational values of oscillation parameters is determined according to the following equation:

$$K_l = \frac{a_r(0) \cdot a_{ef}(0)}{a_r(t) \cdot a_{ef}(t)}, \tag{3.16}$$

where $a_r(0)$ is the initial value of the acceleration oscillation after mounting the bearing, $a_{ef}(0)$ is the initial effective value of the acceleration after mounting the bearing, $a_r(t)$ is the value of the acceleration oscillation during operation, and $a_{ef}(t)$ is the effective value of acceleration during operation.

If the initial values of the oscillation and the effective values of acceleration are unknown, then the reference values $a_r(0) = 40$ ms^{-2} and $a_{ef}(0) = 10$ ms^{-2} are chosen. The extent of bearing damage is then evaluated based on Table 3.3.

3.1.3.2 Analysis of vibrometric signals in the frequency domain

Significantly larger information content about the nature of oscillations and the state of machines can be obtained by *frequency (spectral)* analysis of the time course of deflections, or their speed and acceleration. For this analysis,

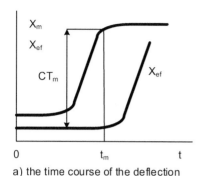
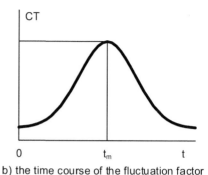

a) the time course of the deflection   b) the time course of the fluctuation factor

Figure 3.13   Time course of integral quantities of oscillation.

Table 3.3   Determination of the extent of bearing damage.

K_l	Bearing condition
0.2 to 1.0	Good bearing
0.05 to 0.2	Slightly damaged bearing
0.02 to 0.05	Damaged bearing
< 0.02	Emergency condition of the bearing

the *discrete Fourier transform* of the DFT of these signals is used, since the digital signal processing takes place only after its time and amplitude discretization by *A/D* converters.

The *Fourier transform* is based on the assumption that every periodic signal $x(t)$ with a period T can be expressed by the sum of harmonic waveforms of frequencies that are determined by a complete multiple of the frequency of the original signal

$$x(t) = \frac{a_0}{2} + \sum_{k=1}^{n} \left[b_k \cdot \sin(k\omega t) + a_k \cdot \cos(k\omega t) \right], \quad (3.17)$$

where n is theoretically infinite, in practice always a finite integer.
The coefficients a_k and b_k of the *Fourier series* are defined by relations

$$a_k = \frac{2}{T} \int_0^T x(t) \cdot \cos(k\omega t) \, dt, \qquad a_k = \frac{2}{T} \int_0^T x(t) \cdot \sin(k\omega t) \, dt, \quad (3.18a, b)$$

where $k = 0, 1, 2, \ldots$ is an integer.

The analyzed signal can also be expressed by *amplitudes* and *phases* of individual spectral components of the signal

$$x(t) = \sum_{k=1}^{n} X_k \cdot \sin(k\omega t + \varphi), \quad (3.19)$$

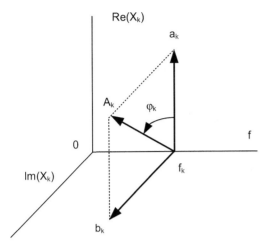

Figure 3.14 Illustration of the frequency component of the analyzed signal.

to which

$$X_k = \sqrt{a_k^2 + b_k^2}, \qquad \varphi_k = \text{actg}\frac{b_k}{a_k}. \qquad (3.20a, b)$$

The *spectral components* of the signal can also be expressed in a *complex form*

$$\hat{X}(f_k) = \frac{1}{T}\int_{-\frac{T}{2}}^{\frac{T}{2}} x(t) \cdot \exp(-j2\pi f_k t dt), \qquad (3.21)$$

where T is the signal period $x(t)$ and $f_k = kf_1 = k/T$ is the frequency of the kth harmonic component of the signal.

To represent the individual spectral components of the analyzed signal, either vector or polar expression is used. The link between these expressions is shown in Figure 3.14.

Knowledge of the *phase frequency spectrum* is particularly suitable for indicating the type of *imbalance of rotating machine parts*, e.g., motor rotors, gears in gearboxes, and bearings.

During the practical processing of the phase spectrum results, it is necessary to determine the so-called *reference point* (*heavy spot*) on the rotating part of the machine (shaft) and place the reference sensor (Keyphasor) there.

Figure 3.15 shows an example of the location of the reference sensor and other sensors on the generator.

Magnetic or optical sensors are used to sense shaft speeds. *Magnetic speed sensors* with telecoil consist of a disc with equidistantly fixed permanent magnets around its perimeter, whose magnetic fields induce voltage

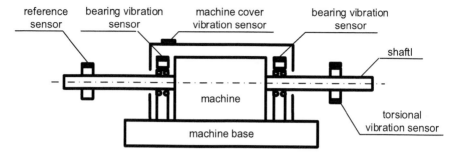

Figure 3.15   Example of the location of vibration sensors on the generator.

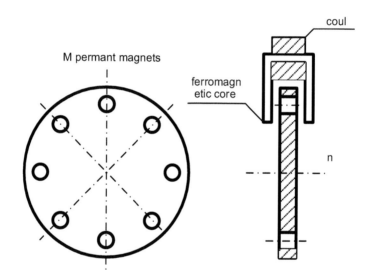

Figure 3.16   Magnetic speed sensor.

pulses in the coil with a frequency determined by the number of permanent magnets *M* and the speed *n* (Figure 3.16)

$$f = m \cdot n. \tag{3.22}$$

Instead of sensing coils, *Hall probes* are also used to indicate changes in the magnetic field of permanent magnets, whose output voltage is proportional to the magnetic field of the magnets.

   Optical speed sensors with a pass-through light beam are formed by an optical system with a light beam source that passes through the holes of the disc and the photodiode, sensing the resulting intermittent optical beam (Figure 3.17a).

   Reflective optical speed sensors use the reflection of a light beam, usually generated by a laser, from evenly distributed reflective surfaces located

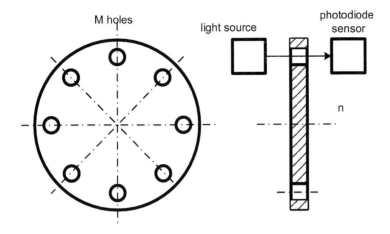

a) through-beam speed sensor

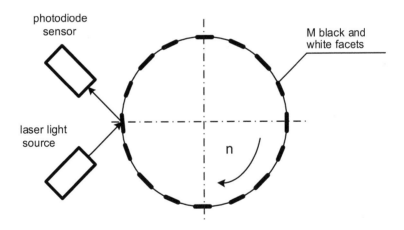

b) reflective speed sensor with zebra tape

Figure 3.17   Optical speed sensors.

on the shaft, e.g., in the form of a *zebra tape* (Figure 3.17b). For both types of optical speed sensors, the pulse frequency is determined by eqn (3.22).

On the frequency spectrum of deflections, we distinguish the area of low and medium frequencies.

The *low frequency range* corresponds to the frequency range of rotational machine parts, evaluating the frequency range of oscillations in the range of several times the number of revolutions of rotating machine parts. Typical sources of *subharmonic* and *interharmonic components* of the frequency spectrum of machines are, for example, dynamic imbalances of their rotating parts, misalignment of their shafts, and hydrodynamic instability of steam generators.

The area of *medium frequencies* corresponds to the frequency range of gear frequencies of gearbox gears. These frequencies are produced when two or more gears are engaged. Each gearbox generates oscillations with a tooth frequency

$$f_z = f_v \cdot fn_v = f_m \cdot n_m, \tag{3.23}$$

where f_v is the rotation frequency of a larger wheel of the number of teeth n_v and f_m is the rotation frequency of the smaller wheel of the number of teeth n_m. The amplitudes of the spectral components of the oscillation rate gradually decrease during the running-in of the gearboxes, but after a certain period of time, determined by the service life of the gears, they increase.

A frequently used parameter to assess the technical condition of gears is the *coefficient of engagement* k_{UF}, defined by the ratio of the smallest common multiple of the number of teeth of both wheels in the N_{SN} engagement to the number of teeth of the larger gear n_v

$$k_{UF} = \frac{N_{SN}}{n_v}. \tag{3.24}$$

The corresponding claw frequency of this phenomenon f_{HT} is

$$f_{HT} = \frac{f_v}{k_{UF}} = \frac{f_v \cdot n_v}{N_{SN}} = \frac{f_m \cdot n_m}{N_{SN}}. \tag{3.25}$$

The spectral component of the f_{HT} frequency is usually low, approximately tenths of Hz, and has a low amplitude, because it occurs only as an oscillation, and only in the case of a damaged gear. Therefore, this frequency component can only be monitored with a sufficiently long recording.

To suppress interfering spectral components in the frequency spectrum of oscillations of rotating machines, which are caused by sources other than gearing, the so-called *kepstral analysis* is often used instead of spectral analysis of signals. This is defined by the absolute value of the *Fourier transform* of the logarithm of the *power spectrum* $\hat{X}(f)^2$ of the analyzed signal

$$C(\tau) = \left| F\left\{ \log \left| \hat{X}(f)^2 \right| \right\} \right|^2, \tag{3.26}$$

where $F\{.\}$ is the symbol of the *Fourier transform*. The variable τ has the dimension of time and is called *kvefrence* (looping of the word frequency).

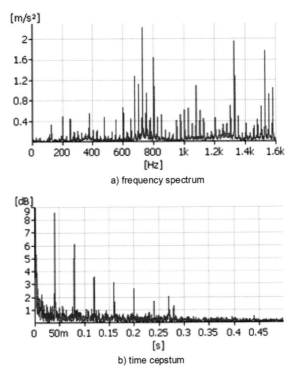

a) frequency spectrum

b) time cepstum

Figure 3.18   Frequency spectrum and time cepstrum acceleration of the defective gearbox.

Cepstral analysis is particularly suitable for the analysis of phenomena whose power spectrum has the character of a product or a proportion of several partial power spectra. After maliciousness, the cepstra can thus be easily determined by the sum or difference of partial spectra. This makes it possible to distinguish the frequency spectrum of a useful signal from the frequency spectra of interfering signals clay of a faulty gearbox.

Figure 3.18 shows the frequency spectrum and cepstrum acceleration of the defective gearbox.

From the above cepstra, it can be found that the original spectrum of acceleration contains two significant sources of periodic components in the spectrum. The first of them has a value of $\tau_1 = 269$ ms (3.714 Hz) and the second more significant source has a value of $\tau_2 = 279$ ms (3.58 Hz), whose components are represented again by a harmonic series in the time kepstra.

Such a small frequency difference is difficult to detect from the normal frequency spectrum. The first source of interference is related to the drive system and the second is caused by the speed frequency of the gearbox input shaft, which is an important source of amplitude modulation, the source of which is, for example, the bending of the gearbox shaft.

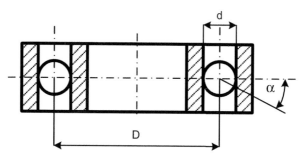

Figure 3.19   Ball bearing dimensions.

The area of *high frequencies* is characterized by defects of rolling bearings. The most common bearing defects are cracks or wells on one of the bearing elements. These defects cause mechanical microrases upon contact with another element of the bearing, which vibrate the bearing at frequencies of several kHz. By demodulation of signals from accelerometers, the kinematic frequency of the bearing can then be determined.

For example, a defect in the outer ring of a bearing in Figure 3.19 with a number of balls n with diameter d, pitch diameter D, and a defect determined by the angle of contact a is manifested by a disturbing signal of frequency

$$f_{ok} = \frac{n}{2} \cdot f_r \cdot \left(1 - \frac{d}{D}\right) \cdot \cos\alpha. \qquad (3.27)$$

Failure on the bearing *inner ring* is manifested by the frequency component

$$f_{ik} = \frac{n}{2} f_r \left(1 + \frac{d}{D}\right) \cos\alpha. \qquad (3.28)$$

where f_r is the relative frequency of rotation of the inner and outer rings of the bearing.

Failure on the *bearing ball* or *roller* generates a disturbing frequency signal

$$f_{ku} = \frac{D}{2n} f_r \left(1 - \left(\frac{d}{D}\cos\alpha\right)^2\right). \qquad (3.29)$$

Bearing cage failure generates a disturbing frequency signal

$$f_{kl} = \frac{1}{2} f_r \left(1 + \frac{d}{D}\right) \cos\alpha. \qquad (3.30)$$

By continuous monitoring of start-ups, operation, and run-up of machines, it is possible to determine trends of gradual deterioration of the technical condition of machines. For this purpose, three-dimensional representations of frequency and time dependencies of velocity or acceleration of oscillation

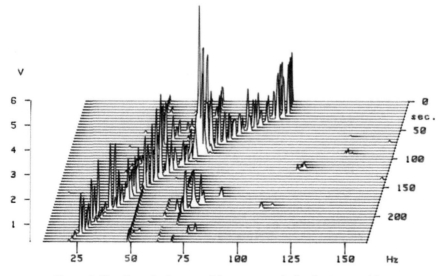

Figure 3.20 Cascade diagram of frequency analysis of rotary machine.

are used. An example of such a *cascade diagram* expressing the time and frequency dependence of the oscillation rate of the rotating machine after its commissioning is given in Figure 3.20.

It is clear from this diagram that the majority frequency of the oscillation rate of this machine is approximately 25 Hz, with the maximum oscillation rate occurring after 100 s from the start of the machine into operation, while the oscillation speed after this time drops below a few tenths of m/s.

3.2 Torsional Oscillation of Shafts

The torsional oscillation of shafts of large rotating machines, e.g., turbogenerators or engines, is caused by changes in the generated or transmitted power and the associated changes in the torque of the shafts.

3.2.1 Causes of shaft torsional oscillation

The source of torque changes in electric rotating machines are torques arising in the air gap of the machine and moments arising from changes in the transmitted power.

The *torque in the air gap* consists of a DC and an alternating component, the frequencies of which correspond to the frequency components of the alternating current produced by the generator or by which the motor is powered.

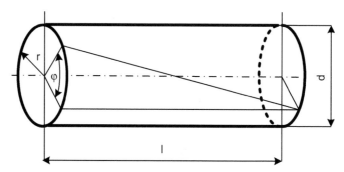

Figure 3.21   Full shaft torsion angle.

The *response moment* occurs in the axial parts of rotary electric machines and is caused by the response to the moment in the air gap. The response moment copies the interference and therefore contains not only the frequency components corresponding to the harmonic components of the electrical network but also the components corresponding to the nominal torsion frequencies of the machine. The response torque, therefore, also depends on the stiffness of the shaft.

If the *circular shaft* of the machine is stressed with torque M_k, then the tangent stress t_k on the shaft surface is defined by the following equation:

$$\tau_k = \frac{M_k}{W_k} \tag{3.31}$$

where W_k is the *cross-sectional modulus in torsion*.

For a *solid shaft* with diameter d and for a hollow shaft with an inner diameter d and an outer diameter D, the modulus of elasticity is defined by the following equations:

$$W_k(d) = \frac{\pi d^3}{16}, \qquad W_k(d,D) = \frac{\pi}{16D}(D^4 - d^4). \tag{3.32a, b}$$

The stress on the shaft causes its *torsion*, which is defined on the basis of *Hook's law*, by the torsion angle

$$\varphi = \frac{M_{\cdot k}}{G.I_p}[\text{rad}], \tag{3.33}$$

where l is the length of the shaft, G is the *modulus of elasticity* of the material in the *shear*, and I_p is the *polar moment of the shaft profile* (Figure 3.21).

For polar moments of circular and intercircular shaft profiles,

$$I_p(d) = \frac{\pi}{32}d^4, \qquad I_p(d) = \frac{\pi}{32}(D^4 - d^4). \tag{3.34a, b}$$

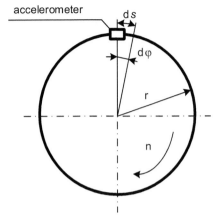

accelerometer

Figure 3.22 Position of the accelerometer on the motor shaft.

The product of $G.I_p$ is called stiffness in torsion.

In addition, the proportional twisting of the shaft is defined, given by its torsion angle j, related to the length l of the shaft, which is expressed either in the arc measure (radians) or in degrees

$$\alpha = \frac{\varphi}{l} = \frac{M_k}{G \cdot I_p}[\text{rad}] = \frac{M_k}{G \cdot I_p}\frac{180}{\pi}\ [^o m^{-1}] \tag{3.35}$$

The permissible value of the proportional torsion of the shafts a is in the range of 0.25–0.35 degrees per 1 m of their length. Its changes, caused by torsional oscillation of shafts, reach several hundredths of a degree per 1 m of shaft length.

3.2.2 Measurement of shaft torsional oscillation

Strain gauge, accelerometric, or optical sensors are used to measure torsional oscillation.

The *strain gauge method* of measuring torsional oscillation evaluates the torsional oscillation of shafts from changes in tangent stress on their surface. For this purpose, *metal foil strain gauges* are used, which change their resistance depending on their length changes. To measure the tangent stress of the shafts, four strain gauges are used, connected to the bridge, while their axes are rotated by 45° and 135°, respectively, with respect to the shaft axis (Figure 3.22).

Bridge connection of strain gauges located at an angle of 45° with respect to the shaft axis eliminates its stress in tension and bending. In addition, the temperature dependence of the resistances of strain gauges is significantly suppressed in their bridge connection.

For the differential stress of the bridge,

$$u_d(t) = \frac{dR}{R} U_n = k_t \frac{dl_t}{l_t} U_n, \tag{3.36}$$

where U_n is the supply voltage of the bridge, and k_t is the strain gauge constant determining the relative change in the resistance of the strain gauge dR/R to its relative change in length dlt/lt. This constant is approximately 2 for *metal foil strain gauges*.

If the shaft is torsionally stressed, then for a proportional change in the length of each of the strain gauges,

$$\frac{dl_t}{l_t} \approx \frac{r^2}{2l_t^2} \varphi, \tag{3.37}$$

where φ is the torsion angle.

On the basis of eqn (3.36) and (3.37), the torsion angle is approximately determined by

$$\varphi \approx \frac{l}{r} \sqrt{2 \frac{u_d}{k_t U_n}} [\text{rad}] \approx \frac{l}{r} \frac{180}{\pi} \sqrt{2 \frac{u_d}{k_t U_n}} [°]. \tag{3.38}$$

Time changes in the differential voltage of the bridges, caused by time changes in the torque, are, after amplification, transferred through the sensing rings to the evaluation unit, or they are transmitted wirelessly by a transmitter with an amplitude or frequency modulated signal to the receiver in which the signal is demodulated and subsequently processed.

The *accelerometric method* of measuring torsional oscillations is based on the measurement of the change in tangent acceleration on the shaft or disc, slid on it, which is triggered by a change in shaft speed. If there is no torsional oscillation of the shaft, the speed is constant, and the tangential acceleration on the shaft surface is zero. However, if the shaft torsional oscillations occur, then there will be a time change in shaft speed dn/dt, which will be directly proportional to the time change in tangent acceleration

$$\frac{da_t}{dt} = r \frac{dn}{dt}, \tag{3.39}$$

where r is the radius of the shaft or disc (Figure 3.23).

By double integration of the signal from the accelerometer located on the shaft, the time change of the circular motion on the shaft surface $ds(t)/dt$ is then determined, which expresses the time change in the angle of its torsion

$$\frac{d\varphi}{dt} = \frac{1}{r} \frac{ds}{dt} [\text{rad}] = \frac{180}{\pi r} \frac{ds}{dt} [°]. \tag{3.40}$$

deformation element   strain gauge bridge

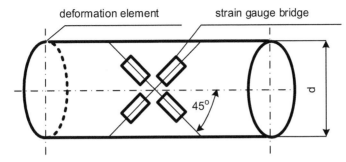

Figure 3.23 Location of strain gauges for torsional vibration detection.

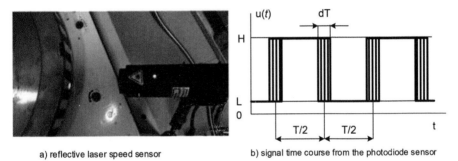

a) reflective laser speed sensor   b) signal time course from the photodiode sensor

Figure 3.24 Measurement of shaft torsional oscillation by reflective optical method.

Similar to the previous method, contact or wireless signal transmission methods are used to transmit signals from accelerometers to the evaluation unit.

Of the *optical methods* of sensing the torsional oscillation of shafts, the most commonly used is the reflective method, which is also used for speed measurement. If there is no torsional oscillation of the shaft, then the pulse signal from the photodiode has stable time parameters. If there is a torsional oscillation of the shaft, then the time changes of these parameters determine the time changes in the shaft torsion angle

$$\frac{d\varphi}{dt} = \frac{1}{r}\frac{dT}{T}[\text{rad}] = \frac{180}{\pi r}\frac{dT}{T}[^{\circ}]. \tag{3.41}$$

Figure 3.24 shows the arrangement of the reflective laser speed sensor with *zebra tape* glued to the machine shaft and the time course of the pulse signal from the photodiode sensor.

Figure 3.25 shows examples of the time course of the electromagnetic moment in the air gap of the steam generator with the corresponding response of changes in the torsional angle of its shaft when changing the load of the generator and turbo-generator with a nominal power of 100 MVA.

M$_k$ [MNm]

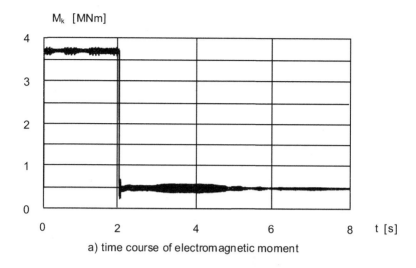

a) time course of electromagnetic moment

φ [°]

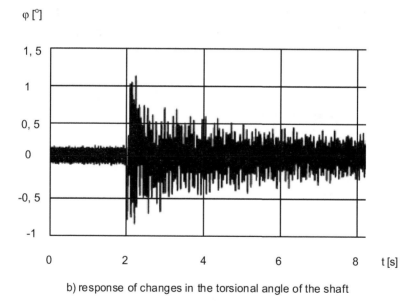

b) response of changes in the torsional angle of the shaft

Figure 3.25   Time courses of electromagnetic moment and changes in torsional angle.

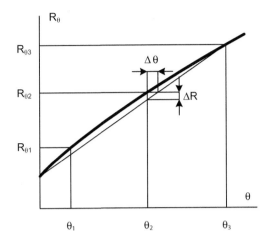

Figure 3.26 Conversion characteristics of the Pt sensor.

3.3 Measurement of Temperature Images of Electrical Machines

Contact and non-contact temperature sensors are used to measure the temperatures not only of bearings but also of insulating systems of large HV electric machines. Thermal cameras are used to measure the temperature fields of electrical machine parts.

3.3.1 Contact temperature sensors

Resistive Pt or Ni metal sensors, PTC, NTC semiconductor sensors, or thermocouples are used for contact temperature measurement.

Platinum temperature sensors have a defined temperature dependence of resistance on temperature by a polynomial in the temperature range of 0°C to + 850°C

$$R_{Pt}(\vartheta) = R_{Pt}(0) \cdot \left(1 + \alpha_{Pt} \cdot \vartheta + \beta_{Pt} \cdot \vartheta^2\right), \tag{3.42}$$

where $R_{Pt}(0)$ is the nominal resistance of the sensor at 0°C, $\alpha_{PT} = 3.91 \cdot 10^{-3}$, and $\beta_{PT} = -5.8 \cdot 10^{-7}$ are the resistance temperature coefficients of the sensor (Figure 3.26).

Platinum sensors are manufactured with nominal resistors ranging from 100 Ω (Pt 100) to 100 kΩ (Pt 1000) in accuracy classes of 0.125°C and 0.25°C.

Nickel temperature sensors have a defined temperature dependence of resistance on temperature by a polynomial in the temperature range of –60°C to 180°C

$$R_{Ni}(\vartheta) = R_{Ni}(20)\left(1 + \alpha_{Ni} \cdot \vartheta + \beta_{Ni} \cdot \vartheta^2\right), \tag{3.43}$$

Figure 3.27   Foil Pt sensor including
its housing.   a) foil Pt sensor   b) Pt sensor housing

where $R_{Ni}(20)$ is the resistance of the sensor at 20°C, $\alpha_{Ni} = 3.83 \cdot 10^{-3}$ and $\beta_{Ni} =$ 4.64·10^{-6}. These sensors are manufactured with nominal resistors $R_{Ni}(20) =$ 100 Ω to 20 kΩ with an accuracy of ± 2°C.

Figure 3.27 shows the foil Pt temperature sensor including its encapsulation, allowing its use in industrial environments.

Polycrystalline PTC (positive temperature coefficient) sensors have a positive temperature coefficient. Their resistance dependence on temperature is defined by the following equation:

$$R_{Si}\left(\vartheta\right) = R_{Si}\left(25\right) \cdot \left(1 + 7.95 \cdot 10^{-3} \cdot \left(\vartheta - 25\right) + 1.95 \cdot 10^{-5} \cdot \left(\vartheta - 25\right)^{2}\right), \quad (3.44)$$

where $R_{Si}(25)$ is the resistance of the sensor at 25°C.

Polycrystalline NTC (negative temperature coefficient) sensors have a negative temperature coefficient. Their resistance dependence on temperature is defined by the following equation:

$$R_{NTC}\left(\Theta_{1}\right) = R_{NTC}\left(\Theta_{2}\right) \cdot \exp\left(B \frac{\Theta_{1}}{\Theta_{2}}\right), \quad (3.45)$$

where $R_{NTC}(\Theta_{1})$ and $R_{NTC}(\Theta_{2})$ are the sensor supports at absolute temperatures Θ_{1} and Θ_{2} and B is the material constant of the sensor, reaching values from 1500 to 7000 K.

To convert the changes in the resistances of these sensors to voltage or current, a connection with instrument amplifiers is used. Figure 3.28 shows an example of connecting a platinum sensor with an *instrument amplifier*.

The platinum sensor is supplied with a current $I_r = U/R_0$ and the voltage drop at its voltage terminals is amplified by an instrument amplifier with paired resistors R_2, R_3, and R_4. For the output voltage of the amplifier

$$U_2 = \frac{R_{Pt}}{R_0} \frac{R_4}{R_3} \left(1 + \frac{R_2}{R_1}\right) U_r + U_0. \quad (3.46)$$

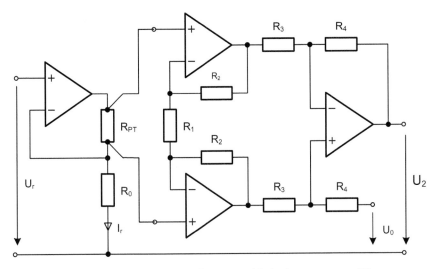

Figure 3.28 Connecting the Pt sensor with the instrument amplifier.

For example, at the reference voltage $U_r = 2.5$ V and $R_o = 2.5$ kΩ, the current flowing through the sensor is 1 mA. For example, if the output voltage of the amplifier is to change in the range of 0–1 V when the sensor temperature changes from 0°C to 100°C, then the difference in gain of the instrument amplifier is 26.3. With identical resistances $R_2 = R_3 = R_4 = 100$ kΩ, the resistance $R_1 = 7.9$ kΩ. Since the voltage drop on the sensor is 100 mV at its temperature of 0°C, the voltage is $U_0 = 2.63$ V.

Monolithic semiconductor temperature sensors use the temperature dependence of the voltage U_{BE} of the BE transition of a *bipolar transistor* connected as a *transistor diode* (Figure 3.29).

The temperature change of the transition junction BE is determined by the following equation:

$$\frac{dU_{BE}}{d\vartheta} = \frac{k}{q_e} \frac{U_{BE}}{U_T} - 2.2 \left[\frac{mV}{K}\right],\tag{3.47}$$

where $k = 1.38 \cdot 10^{-23}$ [J/K] is *Boltzmann's constant*, and $q_e = 1.56 \cdot 10^{-19}$ [C] is the *charge of the electron*.

Thermocouples are used to measure higher temperatures, which are made up of a pair of different metals or their alloys (Figure 3.30).

At different temperatures of the measuring and reference ends of the thermocouple, a thermoelectric voltage is generated in their contact, proportional to the difference in these temperatures

$$U_\vartheta = k_t \left(\vartheta_m - \vartheta_r\right),\tag{3.48}$$

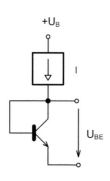

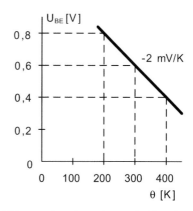

a) connection of transistor diode

b) temperature dependence of voltage UBE

Figure 3.29   Transistor diode.

where k_t is the average voltage constant of the thermocouple over a given temperature range.

Table 3.4 lists the parameters of typical thermocouples.

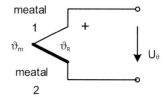

Figure 3.30   Thermocouple.

The effect of the temperature change of the reference end of the thermocouples is eliminated by connecting their reference ends to an *isothermal terminal block* that is located outside the measurement location and whose temperature is measured by semiconductor temperature sensors. Previously, so-called *compensation boxes* were used for this purpose, which contained a bridge connection of temperature-dependent resistors.

Thermocouple amplifier for measurement of the temperature with *isothermal block* is in Figure 3.31.

The transfer of the amplifier, which determines the change of its output voltage with a temperature, is given by the ratio of resistors

$$\frac{R_2}{R_1} = \frac{\Delta U_{OUT} / \Delta \vartheta}{k_t}. \tag{3.49}$$

Temperature dependence of diode voltage U_D with temperature coefficient $dU_D/d\vartheta = -2.2$ mV/°C compensates, if

$$\frac{\alpha_{U_D}}{k_t} = \frac{R_3}{R_1} \tag{3.50}$$

the influence of temperature change of *isothermal block*.

Table 3.4 Parameters of typical thermocouples.

Type	Metal 1	Metal 2	k_t [mV/°C]	Temperature range [°C]
T	Cu	Cu-Ni	39.2	–200 to 400
J	Fe	Cu-Ni	50.2	–200 to 700
E	Cr-Ni	Cu-Ni	60.9	–200 to 1000
K	Cr-Ni	Al-Ni	40.5	–200 to 1300
S	Pt	Pt-10% Rh	10.3	0 to 1500

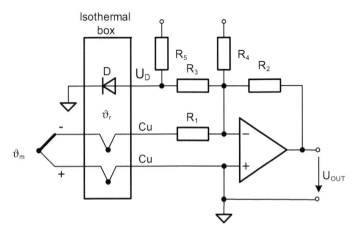

Figure 3.31 Thermocouple amplifier with isothermal terminal block.

The condition of the zero-output voltage of the amplifier at temperature $\vartheta_m = 0°C$ is fulfilled at

$$\frac{U_k}{R_4} = \frac{U_D(0°C)}{R_3}. \tag{3.51}$$

For example, if the output voltage of the amplifier is to be between 0 and 1 V when the temperature of the K-type thermocouple is changed from 0 to 100°C, then when selecting $R_1 = 1$ kΩ, it is $R_2 = 255$ kΩ and $R_3 = 56$ kΩ. At voltage $U_k = -5$ V and voltage on a diode is $U_D(0°C) = 0.689$ V resistance arises is $R_4 = 406$ kΩ. Resistance $R_5 = 4.2$ kΩ determines for a voltage $U_r = 5$ V and temperature 27°C diode current 1 mA.

The *bridge connection of thermocouple voltage amplifier* does not require two opposite polarity supply voltages (Figure 3.32).
Amplifier gains are

$$\frac{R_9}{R_1} = \frac{\Delta U_{OUT}/\Delta\vartheta}{k_t} \qquad \alpha_{U_D} = \frac{R_1}{R_1 + R_2}k_t \qquad U_k\frac{R_4}{R_4 + R_7} = U_D(0°C)\frac{R_1}{R_1 + R_2}$$

$$\tag{3.52a,b,c}$$

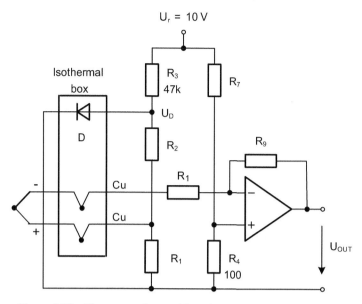

Figure 3.32   Thermocouple amplifier with isothermal terminal block.

Table 3.5   Rated resistors of amplifiers on Figure 3.32.

Types	R_1	R_2	R_7	R_9
K	110 Ω	5.76 kΩ	102.5 kΩ	269 kΩ
J	100 Ω	4.02 kΩ	80.7 kΩ	200 kΩ
S	100 Ω	20.5 kΩ	392 kΩ	1.07 MΩ

Table 3.5 shows the resistors for an output voltage of the amplifier which has a temperature change of 10 mV/°C.

Typical thermocouple signal processing circuits with an internal iso-thermal block are AD594 and AD596 for type J thermocouples and AD595 and AD597 for type K thermocouples. Both circuits achieve temperature measurement accuracy of up to 0.5%. Modern circuits for signal processing of thermocouples are, e.g., AD8496 and AD8497, which are designed for thermocouple types J and K and whose output voltage change is 5 mV/°C. The circuits have temperature compensation for temperature variations in the reference ends of the thermocouples.

When using thermocouple amplifiers, it is also necessary to consider the method of their placement and soldering on the printed circuit boards. In order to eliminate the influence of additional thermoelectric voltage sources, it is necessary to eliminate the temperature differences of the leads at the input of the operational amplifier and to *symmetrize* their geometrical arrangement.

Table 3.6 Emissivity of materials used in the construction of HV electric machines.

Material	Asbestos insulation	Epoxy insulation	PVC	Black steel lacquer	Cast iron	Mild steel	Copper oxidized	Oil paints
Emissivity	0.96	0.89	0.92	0.875	0.44	0.28	0.78	0.94

It is also necessary to select CdSn *low-temperature solder components* in the ratio of 70%–30%, which in conjunction with a *copper conductor* have a thermoelectric constant of 0.3 mV/°C. These solders have thermoelectric constants in order of magnitude less than conventional PbSn solders relative to the copper conductor.

Thermocouples, like resistance temperature sensors, are placed in protective housings with appropriate mechanical, climatic, and electrical resistance.

3.3.2 Non-contact temperature sensors

For non-contact temperature measurement, pyrometers are used, which work on the principle of measuring the energy of radiation produced by any object whose temperature is higher than zero absolute temperature, i.e., −273°C. The radiated energy of the body is determined by the *Stefan–Boltzmann law*, which defines the radiated energy of an *absolute black body* with an area of 1 m² per 1 s

$$E = \varepsilon \cdot \sigma \cdot \theta^4, \tag{3.53}$$

where ε is the *emissivity* of the body whose temperature we measure, θ, is its absolute temperature, and $\sigma = 5.67 \cdot 10^{-8}$ [$Wm^{-2}K^{-4}$] is the *Stefan–Boltzmann constant*.

Table 3.6 lists the emissivity of typical materials used in the construction of HV electric machines.

Calibration of the pyrometer is carried out on a known material at its known temperature.

According to the spectral area used, we distinguish monochromatic, band, and aggregate pyrometers. *Monochrome pyrometers* work with radiation of one or two wavelengths. *Band pyrometers* work with radiation in a wider range of wavelengths. *Aggregate pyrometers* work with radiation in the entire wavelength band.

From the point of view of the optical system, we distinguish *aperture*, *lens*, *mirror*, and *light guide pyrometers*. Photodiodes are used to detect infrared radiation, preceded by lenses with CaF2, ZnS, or ZnSe crystals.

a) portable pyrometer Bosch GIS 1000 C   b) imaging camera GTC 400C

Figure 3.33   Portable pyrometer and thermal imaging camera from Bosch.

To investigate the temperature images of electrical machine parts, terrfo cameras are used, consisting of an optical system and a surface image sensor consisting of a matrix of photodiodes. These sensors use charge-bound *CCD (charge coupled devices)* structures. Their distinguishability, determined by the *number of photodiodes (pixls)*, reaches several tens of thousands to units of millions of pixls.

For example, the portable pyrometer *Bosch GIS 1000 C* with a temperature range of up to 1000°C enables non-contact temperature measurement with a temperature measurement accuracy of 1°C (Figure 3.33a). The thermal imaging camera GTC 400C from the same company with an infrared area sensor with an organization of 120×160 measuring points (pixls), i.e., a total of 19.200 measuring points, allows you to display the temperature field with an accuracy of 5% (Figure 3.33b).

References

[1] Czichos, H. Handbook of Technical Diagnostics. Springer Verlag, Heidlberg, 2020, ISBN: 978-3-642-25849-7
[2] Macusita, A., Tanaka, M., Kanki, H., Kobajasi, M., Keogh, P. Vibrations of Rotating Machinery. Vol. 1 Basic of rotordynamic, practical analysis of diagnostic. Springer Verlag, Tokio, 2017, ISBN: 978-4-431-56658-8
[3] Macusita, A., Tanaka, M., Kanki, H., Kobajasi, M., Keogh, P. Vibrations of Rotating Machinery. Vol. 2 Advanced rotordynamic, application of analysis. Springer Verlag, Tokio, 2019, ISBN: 978-4-431-55452-3

[4] Stein, J., Zayicek, P.: *Steam Turbine-Generator Torsional Vibration Interaction with the Electrical Network*, 2005, ieee.org/document/6345745/

[5] Buskirk, E.: T*orsional Dynamics: Large 2-pole and 4-pole Steam Turbine Powertrains* https://www.gepower.com/content/dam/gepowerpgdp/global/en_US/documents/technical/ger/ger-4724-torsionaldynamics-large-2-and-4- pole-steam-turbine-powertrains.pdf

[6] Václavík, J., Chvojan, J.: *Torsion vibrations monitoring of turbine shafts*, [online], Procedia Structural Integrity, Volume 5, 2017, ISSN 2452-3216. https://www.sciencedirect.com/science/article/pii/S2452321617303153?via%3Dihub

[7] Giesecke, H. D.: Steam turbine-generator torsional response due to interaction with the electrical grid. IEEE Power and Energy Society General Meeting [online]. IEEE, 2012, 2012, s. 1-6 [cit. 2018-04-11]. ISBN 978-1-4673-2729-9. http://ieeexplore.ieee.org/document/6345745/

[8] Bovsunovskij, A., Chernousenko, O., Shtefan, E., Bashta, D.: *Fatigue damage and failure of steam turbine rotors by torsional vibrations strength of materials*, 1/2010, Vol. 42 Issue 1, pp.108-113. http://www.ipp.kiev.ua/jpp-full/2010/2010_01e/108-113.pdf

[9] Halliwell, N.A.: *The laser torsional vibrometer: a step forward in rotating machinery diagnostics*. Journal of Sound and Vibration, Volume 190, Issue 3, 1996, pg. 399-418, ISSN 0022-460X http://www.sciencedirect.com/science/article/pii/S0022460X96900711

[10] Miles, T.J.: *Torsional and bending vibration measurement on rotors using laser technology*, Journal of Sound and Vibration, 226, pp. 441–467. https://www.sciencedirect.com/science/article/pii/S0022460X99922538

[11] *Zebra Tape Butt Joint Correction for Torsional Vibrations*. https://community.plm.automation.siemens.com/t5/TestingKnowledgeBase/Zebra-Tape-Butt-Joint-Correction-for-Torsional-Vibrations/ta-p/354782

[12] ISO 13373-1. *Condition monitoring and diagnostics of machines - Vibration condition monitoring* - Part 1: General procedures. 2002

[13] ISO 10816-3. *Mechanical vibration — Evaluation of machine vibration by measurements on non-rotating parts* — Part 3: Industrial machines with nominal power above 15 kW and nominal speeds between 120 r/min and 15 000 r/min when measured in situ — Amendment 1/2017

4

Insulation Systems of HV Electric Machines

The properties of the insulation systems of HV electric machines, forming their integral part, significantly affect their reliability and serviceability. Insulators must have good mechanical, thermal, and chemical properties in addition to good electrical insulation properties. In terms of state, we distinguish gaseous, liquid, and solid insulators.

4.1 Electrical Properties of Insulators

The basic electrical properties of insulators include their electrical strength, conductivity, dielectric polarization, dielectric losses, and permittivity.

4.1.1 Electrical strength of insulators

The *electrical strength of insulators* is determined by the threshold voltage U_p, which corresponds to the critical value of the electric field E_k at which the insulator breaks through

$$U_p = E_k \cdot d \qquad (4.1)$$

where d is the distance of the electrodes between which the insulator is located. The determination of the threshold voltage is determined not only by the properties and structure of the insulators, but it also largely depends on the size and shape of the electrodes between which the insulator is located.

4.1.1.1 Electrical strength of gaseous insulators
The electrical strength of gaseous insulators is determined by the degree of their ionization.

Ionization of gases is caused by the release of electrons from their atoms, when the originally electrically neutral atom becomes a positive ion. We distinguish between impact, thermal, and photoionization of gases.

Impact ionization occurs by the interaction of force fields during a close passage and deflection of the path of a neutral particle and an electron that has acquired kinetic energy due to the action of an electric field or high temperature

$$W > \frac{1}{2} m_e \cdot v^2,$$ (4.2)

where $m_e = 9.1 \cdot 10^{-31}$ kg is the mass of the electron and v is its velocity.

Photoionization occurs under the action of radiation, the energy of which is determined by *Plank's law*

$$W = h \cdot f,$$ (4.3)

where $h = 6.626472 \cdot 10^{-34}$ [J·s] is *Plank's constant* and f is the frequency of radiation.

Townsend's theory of the avalanche-like mechanism of jump in gases is based on the idea that one free electron, located in an electric field, creates two free electrons and one positive ion. For the number of excited electrons at distance x from the cathode and the number of electrons in the element dx

$$n_x = \alpha.n.x, \qquad dn_x = \alpha.n.dx,$$ (4.4a, b)

where a is the *first ionization coefficient*.

After integrating the previous equations

$$\int_{n_0}^{n_x} \frac{dn_x}{n_x} = \alpha \int_0^x dx,$$ (4.5)

it is possible to derive the relation determining the ratio between the number of excited electrons n_x at distance x from the anode and the number of n_0 in its immediate vicinity (Figure 4.1)

$$\ln\left(\frac{n_x}{n_0}\right) = \alpha.x.$$ (4.6)

This equation can be adjusted to the shape

$$n_x = \cdot e^{\alpha.x},$$ (4.7)

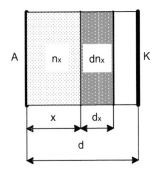

Figure 4.1 Number of excited electrons in a homogeneous electrostatic field.

from which it follows that the current flowing through the gaseous insulator has an exponential course

$$I = I_0 \cdot e^{\alpha . d}. \tag{4.8}$$

where I_0 is the current of the non-ionized gaseous insulator.

A more accurate expression of the processes in the ionized gaseous insulator allows the introduction of a *second Townsend ionization coefficient* γ which additionally takes into account the presence of positive ions and photons that are released when free electrons collide with gas molecules.

If n_0 is the number of electrons released from the cathode and n_+ is the number of electrons released from the cathode after the positive ion hits. The total number of electrons incident on the anode

$$n = (n_0 + n_x) \cdot e^{\alpha . d}. \tag{4.9}$$

The total number of electrons released from the gas is then determined by

$$n - (n_0 + n_+). \tag{4.10}$$

Assuming that each free electron excites one positive ion and that each of these ions releases γ electrons from the cathode, is the total number of electrons released from the cathode

$$n_+ = \gamma[n - (n_0 + n_+)] = \frac{\gamma(n - n_0)}{1 + \gamma}. \tag{4.11}$$

By substituting the number of positive $n+$ ions from the previous equation into eqn (4.9), we determine the total number of free charge carriers in the gaseous insulator

$$n = \frac{n_0 \cdot e^{\alpha . d})}{1 - \gamma(e^{\alpha . d} - 1)}. \tag{4.12}$$

The current flowing through the gaseous insulator is then defined by the dependence

$$I = \frac{I_0 \cdot e^{\alpha \cdot d}}{1 - \gamma\left(e^{\alpha \cdot d} - 1\right)}.$$
(4.13)

The condition of a separate discharge, determining the electrical strength of the insulator, is the unevenness

$$\gamma \cdot e^{\alpha \cdot d} > 1$$
(4.14)

at which the current flowing through the gaseous insulator is close to an infinite value.

Based on *Townsend's theory*, it is possible to deduce the dependence between the jump voltage, pressure, and distance of the electrodes of gaseous insulators at a constant temperature. *Townsend coefficients* α and γ are functionally dependent on the ratio of electric field strength E and pressure p

$$\frac{\alpha}{p} = f_1\left(\frac{E}{p}\right), \qquad \gamma = f_2\left(\frac{E}{p}\right).$$
(4.15a, b)

In a homogeneous electrostatic field with an intensity of

$$E = \frac{U}{d},$$
(4.16)

and after insertion for the ratios α/p and γ/p, it is possible to determine the condition for the formation of a separate discharge in the form of

$$f_2\left(\frac{U}{p \cdot d}\right) \cdot \left(e^{p \cdot d \cdot f_1\left(\frac{U}{p \cdot d}\right)} - 1\right) = 1.$$
(4.17)

Thus, according to *Paschen's law*, there is only one value of stress at a given value of the product of pressure p and the distance of the electrodes d, for which eqn (4.17) applies. This voltage is the threshold voltage

$$U_p = \frac{B \cdot p \cdot d}{\ln\left[\dfrac{A \cdot p \cdot d}{\ln\left(1 + \dfrac{1}{\gamma}\right)}\right]},$$
(4.18)

where the constants A and B are functional dependencies

$$f_1 = A \cdot e^{-\frac{B \cdot p \cdot d}{U_p}}.$$
(4.19)

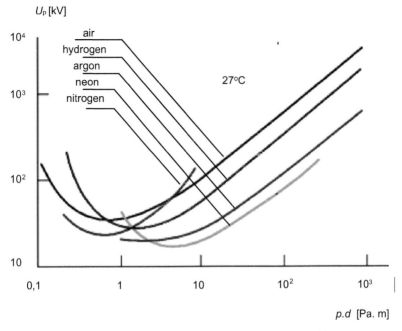

Figure 4.2 Pressure and gas gap dependence of gas threshold stresses.

The threshold voltage acquires its minimum value in the so-called *Stolets point*

$$(p.d)_{min} = \frac{e}{A} \cdot \ln\left(1 + \frac{1}{\gamma}\right). \tag{4.20}$$

Figure 4.2 shows the dependence of the threshold stresses of the most commonly used gases on the product of pressure and distance of the electrodes.

From these dependencies, it is clear that the minimum threshold stresses of gases are achieved approximately at the unit product of their pressure p and the gas gap size d. As the temperature rises, the threshold stresses of the gases decrease approximately linearly with the temperature.

The general condition for the formation of a separate discharge in gases can also be expressed by the *critical concentration* of free electrons at the head of avalanche ionization

$$N_{kr} = e^{\int_0^{x_c < d} \alpha \cdot dx}, \tag{4.21}$$

where x_c is the pathway needed to achieve this concentration. This equation can be rewritten into the form

$$\int_0^{x_c < d} \alpha \cdot dx = \ln(N_{kr}). \tag{4.22}$$

At concentrations of $\ln(N_{kr}) > 18$–20, a spatial charge is formed in the gaseous insulator, which causes a virtually immediate recombination of free charge carriers and the formation of a secondary charge avalanche, into which the remnants of the original avalanche are drawn. This process of so-called proliferation of charge avalanches is referred to as a *stretch mark*.

4.1.1.2 Electrical strength of liquid insulators

Liquid insulators generally have a higher electrical strength than gaseous insulators, but their electrical strength strongly depends on their purity. The electrical strength of liquid insulators depends on many random and almost uncontrollable parameters.

These are, in particular, the presence and type of admixtures such as water and dissolved gases, the degree of contamination of the electrodes, and their shape and the distribution of the electric field between them, the material of the electrodes, the time of action of the electric voltage and its shape course, hydrostatic pressure, the partial pressure of the gases contained in liquid insulators, temperature, and others. Of these causes, affecting the electrical strength, it is difficult to accurately determine the behavior of the liquid insulator under its electrical stress.

Several theories have been developed describing the mechanism of jumping in pure liquid insulators. These theories can be divided into two groups. The first is a group where the jump in liquid insulators causes mainly the emission of electrons from the electrodes and impact ionization. The second group consists of hypotheses in which the formation of air bubbles is a decisive factor.

Electrical discharges and the associated electrical strength of liquid insulators are associated with electrical strength, different nature of the jump mechanism, the value of the jump electrical voltage, the material, and surface treatment of the electrodes. The formation of an electrical discharge usually occurs in places with a higher concentration of impurities. When a spark jumps in the liquid insulator and the penetration of its insulating layer, this damage is not permanent. Therefore, the liquid insulator is always able to regenerate at the point of penetration after disconnecting the electrical voltage.

Figure 4.3 shows the time dependence of the threshold intensity of the electric field of the transformer oil on the time of its action.

From this course, it is clear that the electrical strength of the oil initially reaches up to 40 kV/mm, but after a few tens of microseconds, it already drops to a steady value of approximately 22 kV/mm. Due to contamination of the oil with water and other impurities, gas bubbles occur after this

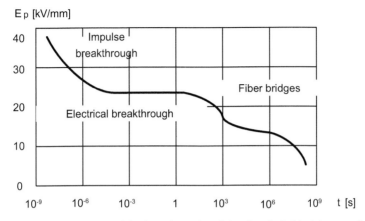

Figure 4.3 Time dependence of the jump intensity of the electric field of the transformer oil.

time, which cause the electrical strength of the oil to drop below the limit of 15 kV/mm.

Without an attached electric field, the liquid is in an electrically disordered state (*unordered state*). The exception is their surface, on which there is a so-called *double layer*, which arises as a result of the lack of bonds of surface atoms or molecules of the surrounding solid insulator at the interface with oil (Figure 4.4a).

When increasing the intensity of the electric field, electrons are *injected* into the liquid. Due to the difference in the structure of the electrode surface and the presence of layers of oxides and impurities, the current density in the liquid insulator can reach up to several kA/mm².

If the intensity of the electric field is already large enough, there is a nonlinear increase in conductivity. The injected electrons create a negative spatial charge around the cathode, which suppresses further electron emission. In the vicinity of the anode, on the other hand, the field is greatly amplified due to the concentration of outgoing electrons (Figure 4.4b).

The injection of charges occurs in places of so-called *microtips* on the surface of the electrode. In these places, channels are formed due to the supplied energy, which heat up and increase their conductivity. Due to the same charge, the channels repel each other and divide into individual bundles of an imaginary cluster (*bunch-like structure*).

In the vicinity of micro-spikes, the electric field is amplified up to hundreds of kV/mm, i.e., more than ten times the macroscopic electrical strength of liquids. In such a strong field, microscopic (*low density volumes*) are formed, which can be considered practically gaseous. According to one

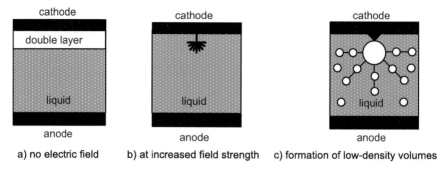

a) no electric field   b) at increased field strength   c) formation of low-density volumes

Figire 4.4   Arrangement of ions in a liquid.

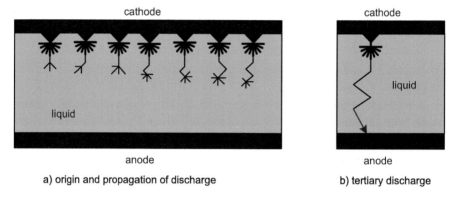

a) origin and propagation of discharge   b) tertiary discharge

Figure 4.5   Formation of strimers and tertiary discharges in liquid insulators.

theory, the *initial process* is initialized within these volumes due to *collision ionization* (Figure 4.4c).

During *impact ionization*, additional low-density volumes occur in the vicinity of the original volume (equivalent to a stretcher avalanche) and the primary streamer is initialized. Inside the resulting volumes, the *secondary streamer* begins to spread. Due to the high current densities, thermal ionization occurs inside the channels, thereby drastically increasing their conductivity. In this case, it is the so-called *leader* (Figure 4.5a).

If the striator reaches the opposite electrode, a *tertiary discharge* occurs inside the ducts. The high current density of this discharge causes oil polarization, light emission, and a drastic drop in the threshold voltage (Figure 4.5b).

We distinguish intrinsic, electrode, volume, and fiber jumps.

Intrinsic breakdown occurs in very pure liquids when they are stressed by impulse voltage and at a small electrode surface. The inception field strength is around 1 MV/mm and has a very low statistical variance.

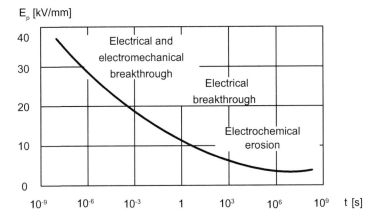

Figure 4.6 Time dependence of the electrical strength of the solid insulator.

Electrode breakdown occurs during short periods of electrical stress on liquids, when drifting of their charged particles does not occur. The development of discharge occurs on the imperfections of the surface of the electrodes. Liquids during this jump no longer exhibit homogeneous behavior and conductive channels are formed in them.

Volume breakdown occurs during prolonged electrical stress, which involves drifting of charged particles. These particles fall on the electrodes and can thus initialize the initial jump process. Thus, the electrical strength of the arrangement depends not only on the volume of the liquid, but also on the distance and dimensions of the electrodes. The electrical strength values in this case show a large statistical variance.

Fiber bridge breakdown is a special case of jumping in oil-cellulose arrangements, such as in transformers. The individual cellulose fibers released are oriented in the direction of the electric field. Fibers often form an unbranched chain between the bridge and the electrodes (*unbranched bridges*). These jumps significantly reduce insulation resistance and electrical strength.

4.1.1.3 Electrical strength of solid insulators

Unlike gaseous and liquid insulators, electrical penetration in solid insulators always causes an irreversible phenomenon. This will result in a loss of insulating ability and permanent burning of the electrically conductive path between the electrodes. Phenomena causing electrical penetration of solid insulating material are divided into three groups in terms of course and character (Figure 4.6).

Purely electric penetration depends only on the size of the threshold voltage, not on the temperature and duration of action of the applied voltage. The energy of the electric field causes an immediately destructive breakthrough at the weakest point of the insulator. The electric strength of the insulator is affected by the homogeneity of the electric field, chemical composition, and material structure and decreases with increasing frequency of the applied voltage.

Thermal breakthrough in solid insulators occurs as a result of their dielectric losses. The heat generated in the insulation system is not enough to dissipate into the surroundings and causes its destruction. Thermal penetration causes a strong dependence of the breakdown voltage on the temperature and duration of action of the attached stress. We distinguish thermal breakthrough under impulse stress and breakthrough in steady state.

Thermal penetration of the solid insulator during its *impulse stress* is caused by a rapid increase in heat in the insulator, and, therefore, its removal to the surroundings is negligible. Thermal conditions of this type of penetration are defined by the equality of specific electrical and thermal energies

$$\sigma E^2 = c_v \frac{d\vartheta}{dt} = c_v \frac{d\vartheta}{dE} \cdot \frac{dE}{dt}, \tag{4.23}$$

where E is the electric field strength, σ is the specific electrical conductivity of the insulator, c_v is its specific heat capacity, and ϑ is its temperature.

Assuming that the specific conductivity of a solid insulator increases exponentially with its temperature

$$\sigma = \sigma_0 e^{-\frac{W}{k \cdot \theta}}, \tag{4.24}$$

where W is the activation energy, $k = 1.38 \cdot 10^{-23}$ [J K^{-1}] is the *Boltzmann constant* and Θ is the *absolute temperature* in K; it can be deduced that the threshold intensity of the electric field causing the penetration of the solid insulator is

$$E_p = \sqrt{\frac{3 \cdot c_v \cdot k \cdot \theta_0^2}{\sigma_0 \cdot W \cdot t_p}} \cdot e^{\frac{W}{2 \cdot k \cdot \theta_0}}, \tag{4.25}$$

where σ_0 is the specific conductivity of the insulator at θ_0, and t_p is the time under impulse stress.

Thermal breakthrough of the insulator at steady state is characterized by the dissipation of thermal energy into the surroundings. Therefore, in eqn (4.23), the term $c_v(d\theta/dt)$ can be neglected, and for the equality of electrical and thermal energy

$$\sigma E^2 = k \frac{d^2 \vartheta}{dx^2}, \tag{4.26}$$

where x is the distance of breakthrough formation from the surface of the insulator.

Assuming that the electric field in the insulator is homogeneous, i.e., that

$$\frac{d^2 U}{dx^2} = 0, \tag{4.27}$$

it can be deduced that the threshold voltage causing the insulator to break through is [4.21]

$$U_p = \sqrt{8 \frac{k}{\sigma} (\vartheta_c - \vartheta_0)}, \tag{4.28}$$

where $\vartheta_c - \vartheta_0$ is the difference in temperature at the breakthrough of the insulator and before its electric field stress.

Electromechanical penetration occurs in solid insulators due to the action of electrostatic forces caused by an electric field. These forces deform the insulation and cause its mechanical instability. If, for example, a plate insulator with a thickness t_0 is stressed by an electric field of intensity $E = U/d_0$, then at steady state, the corresponding force is equal to the deformation force that compresses the insulator to thickness d,

$$\frac{\varepsilon \cdot E^2}{2} = Y \cdot \ln \left(\frac{d_0}{d} \right), \tag{4.29}$$

where ε is the *permittivity of the insulator* and Y is its *Yong modulus of elasticity*.

Mechanical instability of the insulator occurs at a proportional deformation $d/d_0 \approx 0.6$. For the highest intensity of the electric field before the electromechanical breakthrough of the insulator,

$$E_{max} \approx 0.6 \frac{Y}{\varepsilon}. \tag{4.30}$$

Electrochemical breakthrough occurs in cases where an electric field acts on the insulating material for a long time. In the structure of the insulator, electrochemical processes occur, resulting in a decrease in its electrical strength.

4.1.2 Electrical conductivity of insulators

The volumetric electrical conductivity of insulators (*conductivity*), defined by *specific electrical volume conductivity*, also called volume conductivity, is given by the current *I* flowing through the insulator with a cross section of $S = 1$ m^2 and a length $l = 1$ m at a voltage difference of 1 V on its opposite electrodes

$$\gamma_v = \frac{I}{U}\frac{l}{S}\left[S \cdot m^{-1}\right].$$ (4.31)

In the case of solid insulators, we also define the specific *surface conductivity*, determined by the current *I*, flowing on the surface of the insulator with a length of $l = 1$ m at a voltage difference of 1 V on its opposite electrodes

$$\gamma_p = \frac{I}{U}l\left[S \cdot m\right].$$ (4.32)

4.1.2.1 Electrical conductivity of gaseous insulators

The *volumetric electrical conductivity* of gaseous insulators can be expressed by the following equation:

$$\gamma_v = q\sqrt{\frac{g}{r}}\left(b_+ + b_-\right),$$ (4.33)

where $q = 1.6 \cdot 10^{-19}$ [C] is the electric charge of the electron, *g* is the coefficient of its generation, *r* is the recombination coefficient, and *b*+ and *b*– are the mobility of the positive and negative charge carriers.

The dependence of the flowing current through the gaseous insulator and its corresponding volume conductivity on the electric field strength is shown in Figure 4.7.

At a *low electric field strength* of approximately up to 1 kV/mm, this dependence, referred to as the *Ohm's law* region, is approximately linear and the volume conductivity of the gaseous insulator is therefore approximately constant.

In the second region, called the *area of saturated current*, other free charge carriers are no longer ionized, and, therefore, the current flowing through the insulator practically does not change with the increasing intensity of the electric field, and the volume conductivity reaches minimum values, of the order of 10^{-15} [S$\cdot$m^{-1}] to 10^{-17} [S$\cdot$m^{-1}]. With a further increase in the intensity of the electric field to the critical limit E_k, there is an *avalanche-like* ionization of free *charge carriers* (ions) and a significant increase in current, which leads to the breakthrough of the gaseous insulator.

4.1.2.2 Electrical conductivity of liquid insulators

The electrical conductivity of *liquid insulators* strongly depends on their purity. According to the content of impurities, we distinguish extremely pure

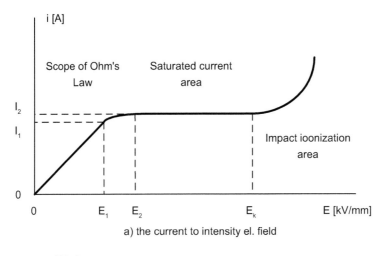

a) the current to intensity el. field

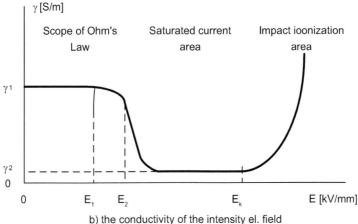

b) the conductivity of the intensity el. field

Figure 4.7 Dependence of gaseous insulator parameters on electric field strength.

liquid insulators and technically pure liquid insulators. In general, the volume conductivity of a liquid insulator can be expressed by the following equation:

$$\gamma_v \approx n \cdot q \cdot b, \tag{4.34}$$

where n is the concentration of free charge carriers (ions) in a liquid with a volume of 1 m³, b is their mobility, and $q = 1.6 \cdot 10^{-19}$ [C] is the charge of the electron.

Extremely pure liquids with perfect degassing show very low conductivity, of the order of 10^{-14} [S·m⁻¹] to 10^{-15} [S·m⁻¹]. Their dependence of the passing current on the intensity of the electric field is similar to that of gaseous insulators. The volume conductivity of technical insulators is either *ionic* or *electrophoretic*.

We distinguish intrinsic and non-improper electrical conductivity. The *inherent electrical conductivity* is due to the dissociation of the molecules of its own liquid insulator. *Improper electrical conductivity* occurs in the case of dissociated admixtures present. To describe the movement of a free ion through a liquid insulator, a *double potential pit* model is used. The probability of an ion jumping between two equilibrium states is determined by the *Boltzmann distribution*. Based on this division, their mobility can be determined

$$b = \frac{f_0 \cdot q \cdot l^2}{6 \cdot k \cdot \theta} \cdot \exp\left(\frac{W}{k \cdot \theta}\right), \tag{4.35}$$

where W [kJ·mol^{-1}] is the energy of the potential barrier that separates the equilibrium positions of ions distant by length l, $k = 1.38 \cdot 10^{23}$ [J·K^{-1}] is *Boltzmann's constant*, Θ is the temperature in K, $f_0 \approx 10^{13}$ [Hz] is the frequency of thermal oscillations, and $q = 1.6 \cdot 10^{-19}$ [C] is the charge of the electron.

By merging the previous equations, it is possible to determine the ratio volume conductivity of liquid insulators

$$\gamma_v = \frac{n \cdot q^2 \cdot f_0 \cdot l^2}{6 \cdot k \cdot \theta} \exp\left(\frac{W}{k \cdot \theta}\right). \tag{4.36}$$

Technically pure liquids achieve specific conductivities of 10^{-11} [S·m^{-1}] to 10^{-12} [S·m^{-1}]. Their dependence of the passing current on the intensity of the electric field is approximately linear up to the critical value of the electric field strength E_k, and, therefore, their conductivity in this area is approximately constant (Figure 4.8).

If the liquid insulator is contaminated with solid particles, it is a *dispersion solution*. If the insulator is contaminated with dissolved liquid substances, it is an *emulsion solution*.

In both cases, the conductivity of the insulating liquids thus polluted is determined by the radius r of spherical colloidal particles, the electrokinetic potential ξ, reaching values of 50–70 mV, the permittivity of the liquid insulator ε and its dynamic viscosity η

$$\gamma_v = \frac{n_k \cdot r \cdot \xi^2 \cdot \varepsilon^2}{6 \cdot \pi \cdot \eta}. \tag{4.37}$$

It follows from these equations that the conductivity of liquid insulators depends on their dynamic viscosity and structure. Therefore, only those liquids that do not have too large dipole moment can be used as liquid insulators.

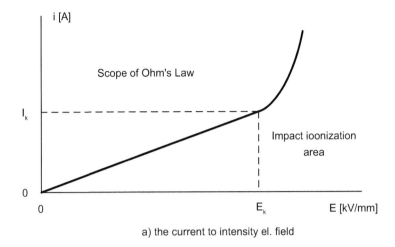

a) the current to intensity el. field

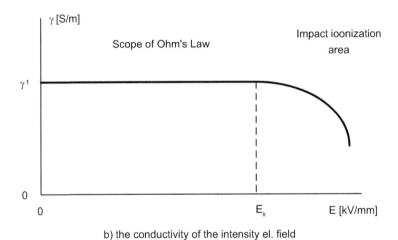

b) the conductivity of the intensity el. field

Figure 4.8 Dependence of technically pure liquides parameters on electric field strength.

4.1.2.3 Electrical conductivity of solid insulators

Solid insulators generally have very little conductivity under normal conditions, i.e., under the action of a weak electric field and low temperatures. It depends on the complexity of the structures of the solid insulator, which are divided into *ionic amorphous* and *crystalline structures*. For solid insulators used in the construction of HV electric machines, solid insulators are used, in which ionic electrical conductivity predominates.

Crystalline substances are made up of positive and negative ions that are firmly held in the nodes of their crystal lattice. Their electrical conductivity can only occur by mutual exchange of positive and negative ions under the condition of the supply of a large amount of energy. Realistically, there are disturbances in the crystals that show some measurable electrical conductivity.

According to *Frenkel's theory*, ions sometimes jump from the lattice node to the inter-node space, the so-called interstitial position, during their thermal oscillations. This results in the creation of just one free space in the structure of the so-called vacancy. This kind of disorder is called *Frenkel's disorder*.

Another kind of disorder, called the *Schottky disorder*, is when there are an equal number of free spots in the crystal lattice after positive and negative ions, and the remaining ions form a new layer of the crystal.

A typical representative of inorganic amorphous substances is, for example, various types of glass. Their electrical but also mechanical properties are mainly influenced by various types of admixtures. The electrical conductivity of different types of glass ranges from 10^{-13} [S · m^{-1}] to 10^{-17} [S · m^{-1}]. The smallest electrical conductivity is shown by *pure silica glass*.

Similarly to liquid insulators, so with solid insulators, their conductivity increases with the increase in temperature, i.e., their resistivity decreases.

In addition to the specific volume conductivity described so far, there is also a specific *surface conductivity for solid insulators*. A significant decrease in surface conductivity causes *surface contamination* of solids by semiconducting or even conductive thin layers.

4.1.3 Polarization of insulators

We distinguish dielectric and relaxation polarization of insulators.

Dielectric polarization is a property of insulators that manifest themselves when they are inserted into an electric field. Under the action of both external and internal electric fields, the electrically bound charges of insulators are moved over small distances from their equilibrium positions and *dipole moments* are thus formed. We distinguish polar and non-polar insulators.

Polar insulators are those isolators in which the centers of their positive and negative charges are not in one place, and these dipoles rotate in the direction of action of the electric field.

Non-polar insulators are those insulators for which there is no inherent dipole moment. The geometric centers of the positive and negative charge carriers are in one place, and after their insertion into the electric field, they deform the molecules in the direction of the electric field intensity. The state of such polarized insulators is expressed by polarization, which is given by the limit ratio of the *dipole moment* to the volume unit of the insulator

$$DM = \lim(\Delta V \to \infty)\frac{\Delta M}{\Delta V}. \tag{4.38}$$

If polarization arises from the displacement of *strongly bound electric charge* carriers, it is a *deformation polarization*, which is always lossless due to the bond strength of the charge carriers.

Relaxation polarization differs from deformation polarization in course time, form, and temperature dependence. Relaxation polarization, unlike deformation polarization, is always lossy. Without the presence of an electric field, the thermal movement of the charge carriers is chaotic, and the probability of the position and distribution of the electric charge is thus the same in all places of the substance.

If we attach an electric field, we change the distribution of potential barriers of charge carriers in the insulator. This creates a *dipole moment* and thus a relaxing polarization. We divide it into ionic and dipole relaxation polarization.

Ion relaxation polarization occurs in insulators, composed of ions or molecules, containing free ions of admixtures and impurities weakly bound in composition to neighboring particles. These insulators are mostly amorphous, such as inorganic glass or ceramic substances containing enamel, and are very sensitive to external electric fields. This kind of polarization can be described by a mathematical model of a *double potential pit*.

It is believed that the charge carriers of both polarities can occupy two different but energetically equal positions. If there is no external electric field, the energy of the two potential pits is the same, the charge carriers overcome the potential barrier due to thermal motion, and these transitions occur with equal probability. Thus, isolation is not polarized. If an external electric field begins to act, there will be a significant change.

The energy of the individual potential pits changes and the particles begin to overcome the potential barrier in places with the lowest energy demands. This leads to an uneven distribution of charge, the formation of a dipole moment, and thus polarization. This polarization is frequency- and temperature-dependent.

Dipole relaxation polarization occurs in polar insulators, which have dipole molecules bound so weakly that they rotate very easily in the direction of field lines due to the action of an electric field. This rotation is prevented by the thermal movement of particles. Therefore, this kind of polarization is very strongly temperature-dependent and also lossy as well as ionic relaxation polarization.

4.1.4 Dielectric losses in insulators

Dielectric losses of insulators occur when they are loaded by an electric field. They are defined by the energy that is converted into thermal energy per unit time in insulators of unit volume. The resulting energy increases the temperature of the insulators and thus increases their specific conductivity and also shortens their service life. This can cause a further increase in dielectric losses and thus a further increase in the temperature of the insulator, which can lead to its thermal breakthrough and destruction.

Dielectric losses can also occur due to *partial discharges* in an unevenly distributed dielectric. They are discharges, occurring only at a fraction of the distance of the electrodes. Discharges have degrading effects on the environment in which they are formed. These are *ionizing losses*, under certain conditions also losses by *partial discharges*.

Dielectric losses can occur due to the action of both DC and AC voltages, regardless of the shape of the voltage curve. Dielectric losses are also temperature-dependent.

In the case of *direct current*, heat *Joule losses* occur, which are proportional to the magnitude of the current and the electrical resistance of the insulator. The *power dissipation* of these losses is determined by the following equation:

$$P_z(ss) = R_i \cdot I^2 = \frac{U^2}{R_i},\qquad(4.39)$$

where R_i is the insulation resistance, U is the voltage attached to the insulator sample, and I is the current flowing through the insulator.

Losses caused by alternating electric fields are given by the sum of thermal ionization and polarization losses. These losses are expressed by equivalent series or parallel loss resistors R_s and R_p in the replacement serial and parallel scheme of the insulator, where C_s and C_p are lossless capacitors (Figures 4.9 and 4.10).

The *dissipation factor* of a *serial replacement* circuit is determined by the ratio of the voltage on its replacement resistor to the voltage on the ideal capacitor

$$tg\delta = \frac{U_{Rs}}{U_{Cs}},\qquad(4.40)$$

For the *AC power dissipation* of the serial replacement circuit,

$$P_{zser}(st) = \frac{\omega \cdot tg\delta \cdot U^2}{1 + tg^2\delta}.\qquad(4.41)$$

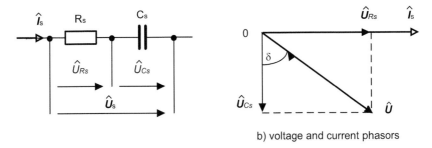

b) voltage and current phasors

Figure 4.9 Serial replacement circuit of the insulator.

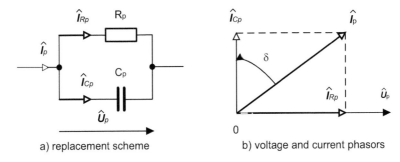

a) replacement scheme

b) voltage and current phasors

Figure 4.10 Parallel replacement circuit of the insulator.

The *loss factor* of a *parallel replacement circuit* is determined by the ratio of current flowing through its replacement resistor to current flowing through the ideal capacitor

$$tg\delta = \frac{I_{Rp}}{I_{Cp}}. \tag{4.42}$$

For the *AC power dissipation* of the parallel replacement circuit,

$$P_{zpar}(st) = \omega \cdot C_p \cdot tg\delta \cdot U^2. \tag{4.43}$$

4.1.4.1 Dielectric losses in gaseous insulators

Dielectric losses of gaseous insulators are mainly due to their electrical conductivity, since polarization losses are negligibly low. For their loss factor, it generally applies

$$tg\delta = \frac{\gamma}{2 \cdot \pi \cdot f \cdot \varepsilon}, \tag{4.44}$$

where γ is the relative conductivity of the gaseous insulator, ε is its permittivity, and f is the frequency.

At a frequency of 50 Hz, the loss factor of gaseous insulators is in the order of 10^{-8} to 10^{-9}.

4.1.4.2 Dielectric losses in liquid insulators

Dielectric losses in liquid insulators depend on the type of polarization of the insulator, the content of impurities that the insulator contains, the temperature, frequency, and intensity of the electric field.

In *pure non-polar liquid insulators*, where only electron polarization is manifested, dielectric losses are determined only by the electrical conductivity of these insulators. Since the electrical conductivity of pure non-polar liquid insulators is negligibly low, dielectric losses are also very small (Figure 4.11).

In the case of *polar liquid insulators*, in addition to losses due to electrical conductivity, polarization losses are also applied. Polar liquid insulators, therefore, exhibit greater dielectric losses than non-polar ones (Figure 4.12).

4.1.4.3 Dielectric losses in solid insulators

In solid insulators, dielectric losses are caused by conductivity, polarization, and ionization losses. Their share in the total losses then depends on the structure of the material, chemical composition, and the presence of impurities.

Non-polar solid insulators contain only lossless elastic polarizations. Dielectric losses are, therefore, small because only conductivity losses are applied in these materials. The dependence of the loss factor on the temperature and frequency of the acting electric field is almost identical to the corresponding dependence of non-polar liquid insulators.

In an alternating electric field, these insulators behave almost like non-polar substances. Dielectric losses are, therefore, mainly conductivity losses, caused by impurities and disturbances of the crystal lattice, and are temperature- and frequency-dependent.

Ion solid insulators with leaking ion deposition have, in addition to elastic polarizations, ionic thermal polarization. Dielectric losses are both conductivity and polarizing in nature. *Ionic amorphous solid insulators* have dielectric losses caused by both conductivity and polarization components.

The content of admixtures, chemical composition, disturbances, and impurities in the structure of the material causes different ratios of the size of the individual components of losses in different substances of this group. Under certain conditions, *ionization losses* can also occur in *cavities filled with gas*.

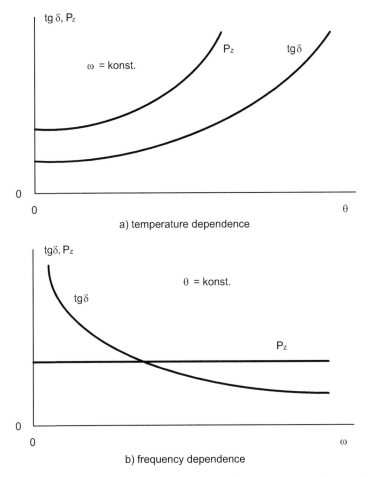

Figure 4.11 Temperature and frequency dependence of the loss factor of non-polar liquid insulators.

4.2 Types of Insulators

According to the state, we distinguish gaseous, liquid, and solid insulators.

Gaseous insulators are further divided into common, rare, and electronegative insulators. *Liquid insulators* are divided into natural and synthetic insulators. Natural insulators are either of vegetable or mineral origin. Synthetic insulators are *silicone oils, polybutenes, chlorinated hydrocarbons,* and *fluorocarbonates.*

Solid insulators are divided into inorganic and organic insulators. *Inorganic insulators* have either an amorphous or crystalline structure.

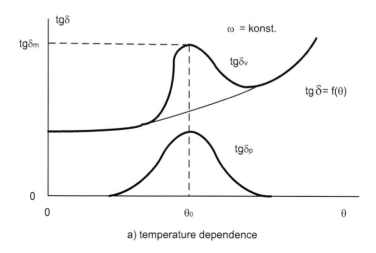

a) temperature dependence

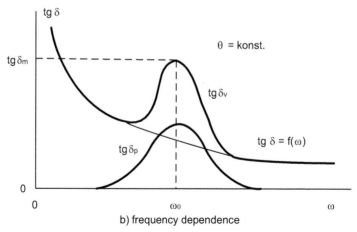

b) frequency dependence

Figure 4.12   Temperature and frequency dependence of the loss factor of polar liquid insulators.

Organic insulators are either of synthetic or natural origin. *Synthetic organic insulators* include *reactoplastics* and *elastomers*.

4.2.1 Gaseous insulators

Gaseous insulating materials have the lowest relative permittivity compared to other insulators. Their biggest advantage over other states is the uniform filling of space and the possibility of use in very tight spaces. They have a very good ability to regenerate after jumping electrical energy. Their dielectric losses are independent of frequency and show very low conductivity values.

The most common gaseous insulators in HV electric machines are air, carbon dioxide, and nitrogen. For example, dry air, consisting of 78.1% nitrogen, 20.9% oxygen, and other gases in neglected quantities, has an electrical strength of approximately 3 kV/mm at a humidity of 50%, a normal pressure of $1.013 \cdot 10^5$ [Pa], and a temperature of 20°C. For example, carbon dioxide CO_2 (2.7 kV/mm), nitrogen N (18 kV/mm), and sulfur hexafluoride SF_6 (9 kV/mm) have more pronounced electrical strength.

Electronegative gases based on chlorine and *fluorine have* a considerable spread. These gases have a reduced mobility of electric charge carriers, which is manifested by their increased electrical strength. The most widely used electronegative gas in the insulation technique is sulfur hexafluoride SF_6, whose relative permittivity is close to one and whose temperature resistance is 150 °C. It is five times heavier than air, and, therefore, it is dangerous to use it in enclosed spaces or trenches and cable shafts.

These gases are used, for example, in encapsulated substations and transformers as an insulating and cooling medium. For its ability to extinguish the electric arc, it is also used in HV switches. In the vicinity of the electric arc, it decomposes, extinguishes the arc, and regenerates again in a fraction of a second. It is also used in coaxial HV cables.

Up to temperatures of 220°C, octafluoro C_3F_8 and dichlorodifluoromethane Cl_2F_2, referred to as freon, are used. All these gases have a significant burden on the environment and are therefore currently being replaced by inert gases or other substitutes.

4.2.2 Liquid insulators

Liquid insulators are used not only as insulation, but mostly also due to their good thermal conductivity, also as a cooling medium, e.g., in HV transformers. They can also be used as impregnations or fillers of solid insulators, where they subsequently improve their electrical properties. According to their origin, they are divided into mineral, vegetable, and synthetic oils.

Mineral oils are a complex chemical mixture of hydrocarbons. They are obtained by distilling crude oil and then refining it. The disadvantage of mineral oils is their aging. Under the influence of external influences (temperature, radiation, absorption of water or moisture from the air, and exposure to oxygen and gases), the electrical properties of these oils change.

The aging of oils can be slowed down by adding inhibitors and regular regeneration. The breakdown voltage of mineral oils is 24–28 kV/mm, the relative permittivity is 2.1–2.4, and the flash point is in the range of 130–150°C.

Figure 4.13 shows the frequency dependence of the loss factor and the relative permittivity of a typical new and degraded mineral transformer oil.

Vegetable oils are obtained from the seeds of plants such as rice, palm, corn, sunflower, rapeseed, etc. Their relative permittivity at a temperature of 90°C ranges from 1.8 to 2.1 and the loss factor reaches 2.10^{-3} to 3.10^{-2} at a frequency of 50 Hz. The advantage of vegetable oils is greater resistance to moisture and the possibility of their ecological disposal compared to mineral oils. The dielectric properties of vegetable oils are significantly dependent on temperature.

Figures 4.13 and 4.14 show the temperature and frequency dependences of the relative permittivity and loss factor of typical transformer and vegetable oils.

Synthetic oils are a suitable substitute for transformer oils. These oils are produced from low molecular weight hydrocarbons, which makes it easy to influence their chemical composition and thus their properties. Synthetic oils are non-flammable and most of them are environmental friendly, are thermally stable, and better resist aging. Synthetic oils include, for example. silicone oils, polybutenes, chlorinated hydrocarbons, and fluorocarbonates.

Silicone oils do not absorb moisture and are chemically inert and non-flammable. Their electrical strength reaches up to 30 kV/mm. They have an average relative permittivity of 2.8 and a loss factor of 5.10^{-3} at a frequency of 50 Hz in the temperature range up to 200°C. Their disadvantage is the considerable dependence of relative permittivity on temperature. Another disadvantage is that they do not absorb gases, such as hydrogen, formed during their aging. This increases the possibility of machine explosion during their use.

Polybutenes form non-polar liquids with very good chemical and electrical properties. Their advantages include excellent insulating properties even at high temperatures and high electrical strength, up to 18 kV/mm. Their typical relative permittivity is 2.2 and the loss factor is 10^{-4} at a temperature of 120°C and a frequency of 50 Hz.

Chlorinated hydrocarbons are not subject to oxidation or aging, even at higher temperatures. They achieve an electrical strength of up to 35 kV/mm. They have a typical relative permittivity of 5.5 and a loss factor of $4 \cdot 10^{-2}$ at a temperature of 150°C and a frequency of 50 Hz. Their disadvantage is that during electrical discharges, hydrogen chloride is formed in them, which corrodes metals and attacks some other insulators.

Fluorocarbonates are non-flammable and explosion-proof compounds. It is a very thermally stable compound that is indifferent to almost all insulations. They achieve an electrical strength of up to 35 kV/mm. Their typical

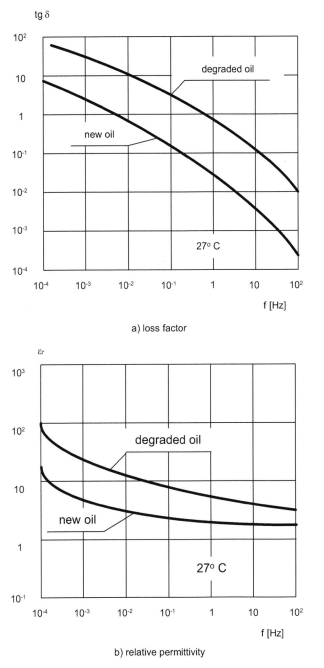

Figure 4.13 Frequency dependence of dielectric properties of transformer oil.

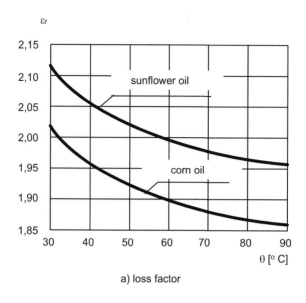

a) loss factor

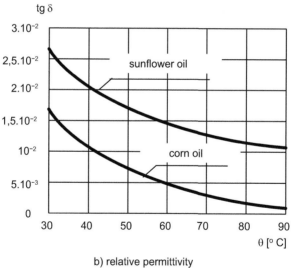

b) relative permittivity

Figure 4.14   Frequency and temperature dependence of dielectric properties of natural oils.

relative permittivity is up to 6 and the loss factor is 10^{-4} at a temperature of 120°C. Table 4.1 shows the electrical parameters of typical liquid insulators.

4.2.3 Solid insulators

Solid insulators are classified according to EN 60085 into the following temperature classes. Class Y (up to 90 °C) includes non-impregnated substances

Table 4.1 Electrical parameters of typical liquid insulators.

Izolant	Electrical strength [kV/mm]	Relative permittivity	Loss factor (50 Hz)	Temperature range
BTA mineral oil	20	2.1–2.4	$1{,}5.10^{-3}$	130°C
Vegetable oils	20	1.8–2.1	$2{\cdot}10^{-3}$ to $3{\cdot}10^{-2}$	90°C
Silicone oils	30	2.8	5.10^{-3}	200°C
Polybutenes	18	2.2	10^{-4}	120°C
Chlorinated hydrocarbons	35	5.5	$4{\cdot}10^{-2}$	150°C
Fluorocarbonates	30	6	10^{-4}	120°C

(cotton, silk, paper, polyvinyl chloride, polyamide fibers, etc.), up to class A (up to 105°C), impregnated fabrics (cotton, silk, paper, cellulose acetate film, volcanofiber, etc.), and up to class E (up to 120°C), hardened cotton fabrics, hardened paper, enameled wires, etc.

Class B (up to 130°C) includes mica, glass fiber, asbestos, and mineral binder pomace, and class F (up to 155°C) includes mica, glass fiber, and asbestos. Class H (up to 180°C) includes silicone elastomers with silicone resin binders. Classes N (up to 200°C) and R (up to 220°C) include mica, porcelain, glass, and quartz in conjunction with inorganic binders.

Of the *macromolecular insulating substances*, polyethylene, polypropylene, polyvinyl chloride, polyamides, and polycarbonates are used to insulate HV electrical machines.

Polyethylene (PE) is an almost non-polar and simplest thermoplastic that is produced by polymerization and currently the most widely used polymer. It has excellent electrical insulating properties that are little dependent on temperature.

Polypropylene (PP) is an almost non-polar thermoplastic polymer that is among the most common plastics. It is a tough, flexible, and hard material. It has excellent electrical insulation properties and good chemical resistance. It is used as a substitute for polyvinyl chloride (PVC) for insulating electrical cables and as insulation in special applications where good ventilation (tunnels) is not possible. Polypropylene produces only limited amounts of smoke and toxic halogen hydrocarbons when burned.

Polyvinyl chloride (PVC) is a dipole substance with molecular-weight-dependent properties. It is currently the most widely used polymer ever. It is a highly chemically and heat-resistant substance, insoluble in water and in a number of organic solvents. It is used as an electrical insulating material and in softened form for insulated cable sheaths.

Polyamides (PI) are among the toughest plastics that are weakly dipole. They exhibit exceptional temperature resistance, up to 500°C for a short time, and are non-flammable, resistant to weather and organic solvents, and resistant to the effects of radiation. One of the important polyamides is an insulating material called *Kapton*. It is used for cable sheathing, groove insulation, flexible printed circuits, wire painting, and impregnation varnishes.

Polyurethanes (PURs) are substances that have very good insulating properties up to 150°C. In electrical engineering, they are used, for example, as encapsulation resins for cable couplings.

Epoxies (EPs) are reactoplastics prepared by polyaddition. In electrical engineering, they are used as casting and casting resins, adhesives, binders, sealants, and paints.

Silicone sealants and *resins* after curing are characterized by considerable heat resistance (up to 200°C, short-term 300°C) and excellent electrical insulation properties. They show very little change in electrical parameters during temperature changes. They are weatherproof and are used as impregnating varnishes or as a binder in glass laminates in the design of engines operating at high temperatures.

Ethylene-propylene elastomers (EPM) have very good resistance to polar agents and are resistant to acids, alcohols, and alkalis. They are used to insulate cables in electrical engineering, where their increased weather resistance is required.

Chlorosulfonated polyethylene (CSM) has very good chemical, thermal, and mechanical properties. They are non-flammable and resistant to oils, ozone, acids, and oxidizing agents. They are used for cable sheaths.

Table 4.2 shows the electrical parameters of the most commonly used solid insulators.

4.3 Electrical Stress on Insulators

The electrical stress of insulators is assessed by the magnitude of the *electric field strength*. When increasing the intensity of the electric field, solid insulators break through, and liquid and gaseous insulators jump. The stress at which a puncture or jump occurs is called a *breakdown voltage*. The intensity of the electric field associated with this voltage is referred to as the *electric strength*.

We distinguish between electrical and thermal breakthrough. *Electrical breakthrough* is caused by sudden ionization of insulator atoms. *Thermal breakthrough* is common for solid insulators with a large loss factor. In this case, the insulator is excessively heated in the electric field, since the heat

Table 4.2 Electrical parameters of typical solid insulators.

Insulant	Electrical strength [kV/mm]	Relative resistivity [Ωm]	Proportional permittivity	Loss factor (50 Hz)	Temperature range [°C]
Impregnated paper (Vulcanfiber)	30	10^6	2.1–2.5	10^{-2}	105
Cable paper	50	10^6	2.2–2.5	10^{-2}	105
Aramide paper (Nomex)	60	10^6	1.9–2.1	$5\cdot10^{-3}$	95
Asbestos insulators	60	10^6	1.8–3.1	10^{-4}	850
Ceramic insulators	30	10^8	6–7	$2\cdot10$	600
Mica insulation	60	10^{13}	2.7–3.2	10^{-3}	850
Glasstext	50	10^8	6–7	$2\cdot10^{-3}$	250
PE linear	42	10^{14}	2.35	$2\cdot10^{-4}$	130
PE branched	38	10^{14}	2.29	$1{,}5\cdot10^{-4}$	130
Polypropylene (PP)	50	10^{11}	2.3	$4\cdot10^{-4}$	130
Polyvinyl chloride PVC	40	10^{13}	3.5	$2\cdot10^{-2}$	240
Polyamides PI (Kapton)	56	10^{15}	3.2	$2\cdot10^{-3}$	500

generated is not enough to dissipate its surface to the surroundings. Thermal breakthrough, unlike electric breakthrough, occurs gradually.

The most common shapes of insulators used in the construction of HV electric machines are plate and cylindrical insulators with homogeneous and layered dielectric.

4.3.1 Stress on plate insulators

Plate insulator with a *homogeneous dielectric* with a thickness d and permittivity e is stressed by a constant electrostatic field, independent of its permittivity of the dielectric. For the capacity of the plate insulator,

$$E = \frac{U}{d}, \qquad C = \varepsilon\frac{S}{d} \qquad (4.45a, b)$$

where S is the area of the dielectric.

Plate insulator with a *layered dielectric* of thicknesses d_1 and d_2 and permittivities ε_1 and ε_2 is stressed by different electrostatic fields,

$$E_1 = \frac{U}{\varepsilon_1\left(\dfrac{d_1}{\varepsilon_1} + \dfrac{d_2}{\varepsilon_2}\right)}, \qquad E_2 = \frac{U}{\varepsilon_2\left(\dfrac{d_1}{\varepsilon_1} + \dfrac{d_2}{\varepsilon_2}\right)} \qquad (4.46a, b)$$

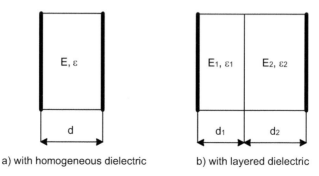

a) with homogeneous dielectric   b) with layered dielectric

Figure 4.15   Plate insulators.

The capacity of a layered plate dielectric is determined by the serial connection of the capacities of its layers

$$C = \frac{C_1 C_2}{C_1 + C_2},$$

(4.47)

where

$$C_1 = \varepsilon_1 \frac{S}{d_1}, \qquad C_2 = \varepsilon_2 \frac{S}{d_2}.$$

(4.48a, b)

The intensity of the electric field of the insulator in this type of insulator is inversely proportional to the permittivity of its dielectrics, i.e., a dielectric with a lower permittivity is stressed with a higher electric field strength than a dielectric with a greater permittivity, since

$$\frac{E_1}{E_2} = \frac{\varepsilon_2}{\varepsilon_1}.$$

(4.49)

Figure 4.15 shows plate insulators with homogeneous and layered dielectric.

4.3.2 Stress on cylindrical insulators

Cylindrical insulator with a *homogeneous dielectric* with an inner radius r_1 and an outer radius r_2 is stressed by a variable electric field strength

$$E_r = \frac{U}{r^2 \left(\dfrac{1}{r_1} - \dfrac{1}{r_2} \right)}.$$

(4.50)

It always reaches the maximum value on the internal electrode of the cylindrical dielectric

$$E_{max} = \frac{U}{r_1^2 \left(\dfrac{1}{r_1} - \dfrac{1}{r_2} \right)}. \tag{4.51}$$

If the difference in radii is significantly less than their mean value, then the intensity of the electric field will be practically constant in a cylindrical homogeneous dielectric

$$E \approx \frac{U}{r_2 - r_1}. \tag{4.52}$$

For the capacity of this cylindrical insulator with a length of l,

$$C = \frac{2 \cdot \pi \cdot \varepsilon}{\ln \dfrac{r_2}{r_1}} l. \tag{4.53}$$

Cylindrical insulator with a *layered dielectric* with radii r_1, r_2, and r_3 and permittivities ε_1 and ε_2 is stressed by variable electric field strengths

$$E_{r1} = \frac{U}{\varepsilon_1 r^2 \left(\dfrac{\dfrac{1}{r_1} - \dfrac{1}{r_2}}{\varepsilon_1} + \dfrac{\dfrac{1}{r_2} - \dfrac{1}{r_3}}{\varepsilon_2} \right)}, \quad E_{r2} = \frac{U}{\varepsilon_2 r^2 \left(\dfrac{\dfrac{1}{r_1} - \dfrac{1}{r_2}}{\varepsilon_1} + \dfrac{\dfrac{1}{r_2} - \dfrac{1}{r_3}}{\varepsilon_2} \right)}. \tag{4.54a, b}$$

The maximum intensity of the electric field in each section of the dielectric is always at its smaller radius. For example, on radius r_1 is the maximum intensity of the electric field

$$E_{1max} = \frac{U}{\varepsilon_1 r_1^2 \left(\dfrac{\dfrac{1}{r_1} - \dfrac{1}{r_2}}{\varepsilon_1} + \dfrac{\dfrac{1}{r_2} - \dfrac{1}{r_3}}{\varepsilon_2} \right)}. \tag{4.55}$$

The electric field strength of this insulator does not depend on the absolute values of permittivities, but only on their ratio

$$\frac{E_{1max}}{E_{2max}} = \frac{\varepsilon_1 r_1^2}{\varepsilon_2 r_2^2}. \tag{4.56}$$

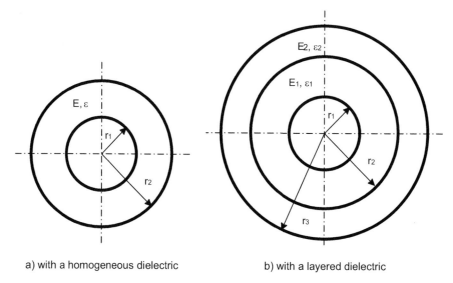

a) with a homogeneous dielectric   b) with a layered dielectric

Figure 4.16   Cylindrical insulators.

The capacity of the cylindrical insulator with length l is determined by the series-connected capacities of the dielectric layers

$$C = \frac{C_1 C_2}{C_1 + C_2},$$
(4.57)

where

$$C_1 = \frac{2 \cdot \pi \cdot \varepsilon_1}{\ln \frac{r_2}{r_1}} l, \qquad C_2 = \frac{2 \cdot \pi \cdot \varepsilon_2}{\ln \frac{r_3}{r_2}} l.$$
(4.58)

Figure 4.16 shows cylindrical insulators with homogeneous and layered dielectric.

It follows from eqn (4.49) and (4.56) that if there are, for example, gas cores in insulators containing gases whose permittivities are always less than those of solid ones, then the surroundings of these cavities are stressed with a significantly higher electric field strength than the solid insulator itself. This phenomenon is the cause of the formation of *internal partial discharges* in both solid and liquid insulators.

4.4 Degradation of Isolators

The electrical insulation properties of insulation systems are most affected by mechanical stress, heat, electric field, and oxidation.

4.4.1 Mechanical aging

Models of mechanical aging of insulators depend on their dielectric proper-ties and various types of mechanical stress. To describe mechanical aging, Oding's *exponential model* and the *Dechtyar–Usipovov model* are most often used.

Oding's *exponential model* is based on the principle of crack growth in the insulator during its tensile stress. These changes move from places subject to greater expansion to places where they are not so stressed by tensile stress.

The *lifetime* of the insulator affected by mechanical aging corresponds to the exponential relationship between the mechanical load and the time to its fracture

$$T_l = \left(\frac{C}{A} \right)^{\frac{1}{m}} \cdot \exp\left(-\frac{\alpha + \beta}{m} \right) \cdot \sigma, \qquad (4.59)$$

where σ is the mechanical stress of the insulator, m is an indicator of the abil-ity to change it, and A, C is α, β are the material constants.

The *Dechtyar–Usipovov model* is based on the idea of the role of changes in the insulator, as a result of the physical disruption of the insulator, which is similar to the germs of its melting. By disrupting the insulator, a certain critical mass of germs of its liquid phase is formed as a result of an increase in the concentration of melting foci. The result of the model is the lifetime of the electrical insulation system, which is an exponential depen-dence of the voltage on the time to refraction

$$T_l = B \cdot \exp\left(\frac{Q_0}{R \cdot \Theta} \right) \cdot \exp\left(-\frac{q \cdot V_a \cdot \sigma}{2 \cdot R \cdot \Theta} \right), \qquad (4.60)$$

where B is the material constant, V is the volume of the mole of atoms, q is the stress concentration coefficient in the insulator area being violated, Q_0 is the activation energy of self-diffusion, Θ is the temperature in K, s is the mechanical stress, and $R = N_a \cdot k = 8.13$ [J·K^{-1}] is the universal gas constant, given by the product of *Avogadro's constant* $N_a = 6.022 \cdot 10^{23}$ [mol^{-1}] with *Boltzmann's constant* $k = 1.38 . 10^{-23}$ [J·K^{-1}].

4.4.2 Thermal aging

Significant thermal aging of insulators occurs due to their oxidation, thermal stress, and chemical decomposition.

Thermal stress on *solid insulators* can manifest itself either globally in the entire volume or locally at a specific location. Local temperature increases are caused, for example, by partial discharges in oils, e.g., on the edges of conductors.

Thermal stress on *liquid insulators* (oils) causes irreversible chemical processes in them, such as oxidation, polymerization, polycondensation, decomposition of some components, depolymerization, and cross-linking of macromolecules. With increasing temperature, the speed of these degradation processes increases. These processes worsen their electrical parameters, e.g., loss factor.

In thermal aging, the main role is played by the amount of the so-called *activation energy*, which is supplied to the insulators so that a reaction can take place, resulting in its degradation. The higher the insulator has this energy, the more it resists degradation processes in its internal structure. When this energy limit is overcome, the isolator particles gain energy to leave their stable state and thus cause the initiation of degradation processes. The rate of degradation reaction is defined by the *Arrhenni equation*

$$k_d = f \cdot \exp\left(-\frac{E}{R \cdot \Theta}\right), \tag{4.61}$$

where k_d is the degradation constant, f is the frequency of collision of oil molecules, E is the activation energy, $R = 8.314$ [$K^{-1} \cdot mol^{-1}$] is the universal gas constant given by the product of *Avogadro's constant N_A* and *Boltzmann's constant k*, and Θ is the absolute temperature.

The following models are used to express the properties of thermal aging.

Montsiger's model assumes that the time of failure, determining the life of the insulator, can be expressed by exponential dependence

$$L_T = A \cdot \exp(-m \cdot \Theta), \tag{4.62}$$

where A and m are material constants.

Assuming that for the constant m

$$m = \frac{\ln 2}{\Delta \Theta}, \tag{4.63}$$

Based on the previous equations, the expected service life of the insulator can be determined

$$L_T = 7.15 \cdot 10^4 \cdot \exp(-0.086 \cdot \Theta), \tag{4.64}$$

For example, if the temperature of the insulator rises by 8°C, the service life of the insulator is reduced to one-half, and, conversely, if the temperature is reduced by 8°C, this service life is doubled.

Dakin's model assumes that the degradation process is of chemical origin and rising temperature accelerates the rate of degradation. This assumption is expressed by the so-called Arrhein equation

$$N = N' \cdot \exp\left(\frac{\Delta W_A}{k \cdot \Theta}\right),$$ (4.65)

where N' is the frequency of collisions of the reacting molecules, DW_A is the activation energy of the main reaction involved, k is the *Boltzmann constant*, and Θ is the absolute temperature.

This equation establishes the time limit from which insulation is considered unusable

$$L_T = \frac{C}{N} = A \cdot \exp\left(\frac{B}{\Theta}\right),$$ (4.66)

where C is the proportionality constant and for the constants A and B

$$A = \frac{C}{N'}, \qquad B = \frac{\Delta W_a}{k}.$$ (4.67a, b)

Eqn (4.66) is commonly displayed in the so-called equation. *Arrheinn's diagram*, which is the dependence of $\log(L_T)$ on the negative inverse value of the absolute temperature $-1/\Theta$. This will allow the results at the test temperature to be extrapolated to the normal operating temperature. International standards have introduced the TI temperature index, which determines the heat resistance of insulating materials. In technical practice, eqn (4.66) is considered as the basis for calculating the lifetime of insulating materials.

Büssing's model is based on the *Monsinger* and *Arrheinn's models*, with the difference that the reaction rate is not constant and depends on the instantaneous concentration of the reacting components

$$\frac{dc}{dt} = -k_d \cdot c^n,$$ (4.68)

where c is the concentration of the reacting components, k_d is the degradation constant, and n is the order of reaction, taking values from 1 to 3.

The service life of the insulator is then determined by the following equation:

$$L_T = B \cdot \exp\left(\frac{B}{\Theta}\right),$$ (4.69)

where material constants b and B can be determined by physico-chemical calculations based on chemical reaction kinetics.

4.4.3 Electrical aging

Since electrical insulating fluids are under constant exposure to the electric field, electric aging is one of the main factors that affect the parameters and quality of insulating fluids. Under the action of the electric field, the dielectric losses of the insulator increase, and, thus, its stress occurs. At higher electric field strengths, a puncture can occur at the point of local stress.

Two different models are used to describe electrical aging.

The *power model* defines the insulation life by equations

$$L_m = C_1 \cdot E^{-n}, \tag{4.70}$$

where E is the electric field strength and C_1, and n is the material constants.

The *exponential model* defines the insulation life by equations

$$L_e = C_E \cdot \exp(-h \cdot E), \tag{4.71}$$

where C_E and h are material constants.

Figure 4.17 shows the dependencies of the insulation life on the electric field strength for the power and exponential models.

In addition, all these material constants are temperature-dependent. The constants n and h represent the coefficients of electrical resistance and are considered the basic parameters determining the insulating properties of insulators. The disadvantage of these models is that they do not take into account the presence of partial discharges and that they lack validity at low electric field strengths, since in such cases, eqn (4.70) and (4.71) would determine the limit infinite lifetime of the insulators.

The lifetime threshold is defined by the following equation:

$$L_E = L_0 \cdot \left(\frac{E}{E_0} \right)^{-n}, \tag{4.72}$$

where L_0 is the service life of the insulation without electrical stress and E_0 is the intensity of the electric field at which degenerative processes do not take place.

When using an exponential insulation life model, the insulator threshold

$$L_E = C_E \cdot \frac{\exp(-h \cdot E)}{E - E_t}, \tag{4.73}$$

where E_t is the *threshold of electrical stress* from which degradation processes begin.

Based on eqn (4.72) and (4.73), it is possible to derive the relation determining the ratio of the insulation lifetime under L_E electrical stress to the life

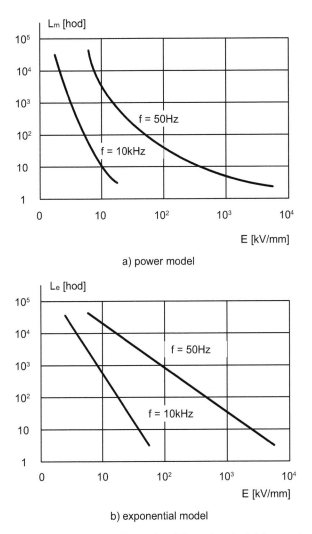

Figure 4.17 Model of insulation life on electric field strength.

of insulation L_0 at an electric field strength in which degenerative processes do not take place

$$\frac{L_E}{L_E} = \left(\frac{E - E_t}{E_t - E_0} \right)^{-n}.$$ (4.74)

This interpretation of the threshold limit of electrical stress of insulators makes it possible to more credibly determine their service life even at low electric field strengths.

4.4.4 Combined aging

In the case when the insulation system is exposed to multiple aging factors at the same time, interactions occur between them. These factors can act either separately or together. The most well-known models describing the combined aging of insulation are Simoni, Ramu, Fallou, and Crine models.

Simoni's model defines aging insulation by the following equations:

$$R = A \cdot \exp\left(-\frac{B}{\Theta}\right) \cdot \exp\left[\left(a + \frac{b}{\Theta}\right) \cdot \ln\frac{E}{E_0}\right], \tag{4.75}$$

where E_0 is the intensity of the electric field at which degenerative processes do not take place.

Combined thermal and electrical stress according to *Simoni's model* will cause reaction rate

$$R(T,E) = R_0 \cdot \exp\left[B \cdot \Delta\left(\frac{1}{\Theta}\right)\right] \cdot \left(\frac{E}{E_0}\right)^{\left[n - b \cdot \Delta\left(\frac{1}{\Theta}\right)\right]}, \tag{4.76}$$

where

$$R_0 = A \cdot \exp\left(-\frac{B}{\Theta_0}\right), \qquad n = a + \frac{b}{\Theta_0}, \qquad \Delta\left(\frac{1}{\Theta}\right) = \frac{1}{\Theta} - \frac{1}{\Theta_0}$$

$$\tag{4.77a, b, c}$$

and Θ_0 is the reference absolute temperature, e.g., 300 K, i.e., 27°C.

Based on the above equations, it is possible to determine the so-called *general equation of the service life of insulation during its combined aging*

$$L(T,E) = L_0 \cdot \exp\left[-B \cdot \Delta\left(\frac{1}{\Theta}\right)\right] \cdot \left(\frac{E}{E_0}\right)^N, \tag{4.78}$$

where

$$N = n - b \cdot \Delta\left(\frac{1}{\Theta}\right). \tag{4.79}$$

This insulation life equation can be displayed as a surface that shows the dependence of the insulation life L under simultaneous exposure to temperature and electric field strength E (Figure 4.18).

The equation of the service life of insulation under combined stress implies that the service life of insulation is determined by the product of the service life at its thermal and electrical stress

$$\frac{L(T,E)}{L_0} = \frac{L(T)}{L_0} \cdot \frac{L(E)}{L_0} \cdot \left(\frac{E}{E_0}\right)^{\left[b \cdot \Delta\left(\frac{1}{\Theta}\right)\right]}. \tag{4.80}$$

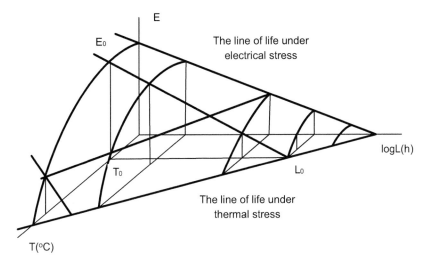

Figure 4.18 Illustration of the service life of insulation under its combined electrical and thermal stress.

The *Ramu model* defines the aging of an insulation by the product of the factors of the individual models of its stress, considering their synergy

$$L(T,E) = c(\Theta) \cdot E^{-n(\Theta)} \cdot \exp\left[-B \cdot \Delta\left(\frac{1}{\Theta}\right)\right], \tag{4.81}$$

where

$$c(\Theta) = \left[c_1 - c_2 \cdot \Delta\left(\frac{1}{\Theta}\right)\right], \qquad n(\Theta) = n_1 - n_2 \cdot \Delta\left(\frac{1}{\Theta}\right), \tag{4.82a, b}$$

and c_1, c_2, n_1, and n_2 are material constants.

Using the relation for $\Delta(1/\Theta)$, Ramu's model can be expressed in the following form:

$$L(T,E) = c(\Theta) \cdot E^{-\left[n_1 - \cdot \Delta\left(\frac{1}{\Theta}\right)\right]} \cdot \exp\left[-B \cdot \Delta\left(\frac{1}{\Theta}\right)\right], \tag{4.83}$$

where c and B are material constants.

If the electrical stress threshold E_0 is part of this equation and the constant is eliminated by substituting after $L = L_0$, $E = E_0$ and $\Theta = \Theta_0$, then *Raunnu's model* is identical to *Simoni's model*.

Fallou's model exploits the exponential dependence of aging insulators

$$L = \exp\left[A(E) + \frac{B(E)}{\Theta}\right], \tag{4.84}$$

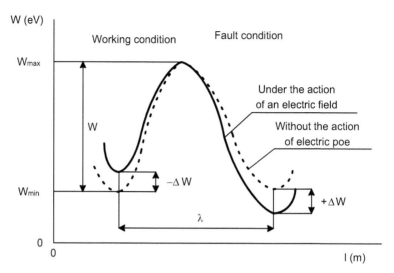

Figure 4.19   Crine's aging model using a double potential pit.

where $E > 0$, $A(E) = A_1 + A_2 \cdot E$, and $B(E) = B_1 + B_2 \cdot E$.

The courses of functions $A(E)$ and $B(E)$ are experimentally determined from the waveforms of the failure times of the insulator at constant temperature. After inserting these waveforms into the previous equation, we obtain the relation determining the life of the insulator

$$L = \exp A_1 \cdot \exp\left(\frac{B_2}{\Theta}\right) \cdot \exp\left[-\left(A_2 + \frac{B_1}{\Theta}\right) \cdot E\right]. \tag{4.85}$$

Crine's model is based on the physical theory of aging. The aging process is explained using a *double potential pit* (Figure 4.19).

In order for the insulator to go from a working state to a state of failure, it needs to overcome the energy barrier that separates its working state from the state of failure. The probability of obtaining the necessary energy to overcome the energy barrier is determined by the *Maxwell–Boltzmann distribution*. In this model, an essential role is played by the action of an electric field, which significantly deforms the energy barrier.

Under the action of the electric field energy W, the height of the barrier on the side of the operating state is reduced by ΔW. On the part of the fault state, on the contrary, it increases by the same value. For this reason, the electric field greatly increases the likelihood of overcoming the energy barrier. By analogy with other models, the time required to overcome the energy barrier is defined as the lifetime of a material.

The *Crine model* uses the laws of thermodynamics to determine life expectancy using the *Maxwell–Boltzmann distribution*

$$L = \left(\frac{h}{k \cdot \Theta}\right) \cdot \exp\left(\frac{\Delta}{k \cdot \Theta}\right) \cdot \cosh\left(\frac{q \cdot \lambda \cdot E}{k \cdot \Theta}\right), \tag{4.86}$$

where $h = 6.62607015 \cdot 10^{-34}$ [J·s] is the *Plank's constant*, $k = 1.38 \cdot 10^{-23}$ [J·K^{-1}] is the *Boltzmann's constant*, ΔW is the free activation energy, λ is the distance between the two states, and q indicates the electric charge of the particles affecting aging.

Crine's model ceases to apply if an electric field is not present and therefore loses use. The parameters ΔW and λ are temperature-dependent and are not precisely defined. Determining these parameters is difficult and depends on the given conditions of electrical stress.

At high electric field strengths, $q \cdot \lambda \cdot E < k \cdot \Theta$ and the service life of the insulator is

$$L = \left(\frac{h}{k \cdot \Theta}\right) \exp\left(\frac{\Delta W - q \cdot \lambda \cdot E}{k \cdot \Theta}\right). \tag{4.87}$$

At extremely high values of electric field strength, the life curve of insulators has an exponential course, since $\lambda \cdot E \ll k \cdot \Theta$. Conversely, at low electric field strengths, Crine's model approximates Simoni's model with threshold values, because the life curve of insulators has a gradually decreasing course.

When an *AC electric field* is applied to insulators, their service life is defined by the following equations:

$$L = \left(\frac{h}{2 \cdot f \cdot k \cdot \Theta}\right) \cdot \exp\left(\frac{\Delta W}{k \cdot \Theta}\right) \cdot \cosh\left(\frac{\varepsilon}{2} \frac{\Delta W}{k \cdot \Theta} \cdot \frac{\Delta V \cdot E^2}{k \cdot \Theta}\right), \tag{4.88}$$

where ΔV is the activation level of the process under electrical stress, f is the acceleration coefficient, proportional to the frequency of the AC electric field, and e is the permittivity of the insulator.

References

[1] Naidu, M.S. *High voltage engineering*, 2nd. ed. New York: McGraw-Hill, ISBN: 007-4622-862
[2] Chpudhury, P.K. *Dielectric materials and application.* New York: McGraw-Hill. ISBN: 978-1536-1531-63
[3] Stone, G.: *Electrical insulation for rotating machines: design, evaluation, aging, testing, and repair.* Hoboken, NJ: Wiley-Interscience, 2004. IEEE Press series on power engineering.

[4]   Pyrhonen, J., Jokinen, T., Hrabovcová, V. *Design of rotating electrical machines*. Second edition. Chichester, West Sussex, United Kingdom: Wiley, 2014. ISBN: 9781118701621

[5]   *Dielectric loss*. http://encyclopedia2.thefreedictionary.com/Dielectric+Loss.

[6]   *Electrical properties of insulating materials*. http://www.readorrefer.in/article/Electrical-properties-ofinsulating-materials

[7]   TEMCo.   http://www.temcoindustrialpower.com/product_selection.html?p=insulation_varnish_di p_and_bake

[8]   *Recommended Practice for Testing Insulation Resistance of Electric Machinery*. New York: The Institute of Electrical and Electronics Engineers, 2014.

[9]   EN IEC 60034-27-4: *Electric rotaring maschienes-* part 27-4. *Measuring of resistance isolation and index polarization in HV electrical maschines*. 2018.

[10] EN 60085 *Elektric isolation - Termerature evaluation and marking*. 2. Prague, 2008.

[11] Nomex 410. Dupont.   http://www.dupont.com/content/dam/assets/products-and-services/electronic-electrical- 58

[12] Nomex 411. Dupont.   http://www.dupont.com/content/dam/assets/products-and-services/electronic-electricalmaterials/assets/411.pdf

[13] Nomex 418. Dupont http://www2.dupont.com/Energy_Solutions/en_US/assets/ downloads/418_419.pdf

[14] Kapton HN. Dupont.   http://www.dupont.com/content/dam/dupont/products-and-services/membranes-andfilms/polyimde-films/documents/DEC-Kapton-general-specs.pdf

[15] Myoflex   https://www.fabrico.com/sites/default/files/datasheets/Myoflex_Flexible_ Laminates_Ed ition_2013.pdf

5

Measurement of Properties of Insulation Systems of HV Electrical Machines

When diagnosing the condition of HV insulation systems of electrical machines, their insulation current, resistance, capacity, loss, and absorption factor are most often determined.

5.1 Measurement of Insulation Current and Resistance

Insulation current is a DC current flowing through the insulation system to ground. When measuring this current, the insulation system of the machine is connected to the HV source of the test voltage U_z and the flowing insulation current I_{iz} is measured by a DC microammeter.

Insulation resistance is determined from Ohm's law

$$R_{iz} = \frac{U_z}{I_{iz}}. \tag{5.1}$$

The *recommended voltage levels* for the DC test voltages at which the insulation resistances are measured are 500 V, 1 kV, 2.5 kV, 5 kV, 10 kV, and 15 kV.

Figure 5.1 shows the voltage dependence of the insulation resistance of the HV standing insulation system with a nominal power of 15 MVA and a nominal voltage of 6 kV. It is clear from the figure that the intact insulation has an insulation resistance virtually independent of the test voltage. Conversely, when the insulation is broken, the insulation current increases with increasing voltage and the insulation resistance decreases until there is an irreversible damage (breakthrough) of the insulation.

In practice, the change of the insulation resistance of the HV insulation system of the machine at its test voltage is evaluated either by its *relative voltage* or by *a time change*.

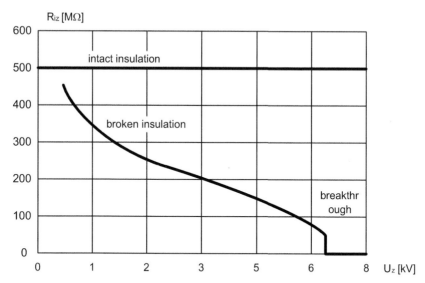

Figure 5.1   Voltage dependence of the insulation resistance of the HV winding of an electric machine.

The *relative voltage change* of the insulation resistance is expressed by the following equation:

$$\delta_{iz} \cdot R = R_{iz0} \frac{di_{iz}}{du_{iz}},\qquad(5.2)$$

where R_{iz0} is the *initial value* of the insulation resistance, corresponding to the linear dependence of the insulation current on the test voltage, and di_z/du_{iz} is the relative change in the conductivity of the machine's insulation system. In the area of linear dependence of insulation current i_{iz} on test voltage U_z, the relative change of *insulation resistance* $\delta R_{iz} = 1$. If the change of insulation current is greater than the change of test voltage, then the relative change of insulation resistance is $\delta R_{iz} > 1$. In practice, it is verified that at $\delta R_{iz} > 5$, there are irreversible changes in the insulation, which can lead to its destruction.

Figure 5.2 shows the dependences of the insulation current and the relative changes in the conductivity of the HV insulation system of the electric machine.

The *time change of the insulation resistance* is defined by the ratio of the insulation time constant $R_i \cdot C$ to the relative change of the insulation resistance $\delta_{iz} R$

$$\tau_{iz} = \frac{R_{iz0} \cdot C}{R\delta_{iz}},\qquad(5.3)$$

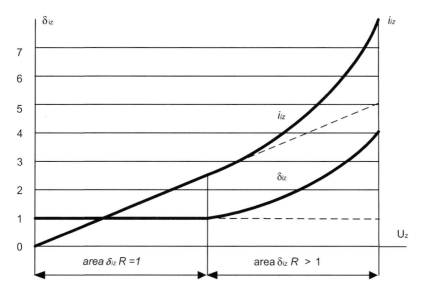

Figure 5.2 Dependence of the insulation current and the change of the insulation resistance on the test voltage.

where C is the capacity of the machine insulation system.

This time constant is used to estimate the aging rate of insulation according to the standard EN 60 505: *Evaluation and classification of electrical insulating materials 2005*.

When measuring the insulation current and resistance, it must be ensured that the measured results are not distorted by additional *leakage currents*, caused, for example, by imperfect grounding of the machine, its surface contamination, or the occurrence of a direct current electromagnetic field approximately the measured machine.

5.2 Measurement of Capacity and Loss Factor

The measurement of these quantities is based on the principle of measuring real and imaginary components of currents flowing through the insulation system of electrical machines when they are stressed by alternating voltage.

If we start from the series replacement scheme of the insulation system in Figure 5.3, where R_s is the *series loss resistance* and C_s is the *series insulation capacity*, then for impedance $\hat{Z}_s$, admittance $\hat{Y}_s$, and *loss factor* $tg\delta$ of the insulation system

$$\hat{Z}_s = \frac{1}{\hat{Y}_s} = \frac{\hat{U}_s}{\hat{I}_s} = R_s + \frac{1}{j \cdot \omega \cdot C_s}, \qquad tg\delta = \frac{\hat{U}_{Rs}}{\hat{U}_{Cs}} = \omega \cdot R_s \cdot C_s. \qquad (5.4a,b)$$

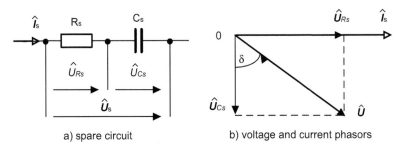

a) spare circuit   b) voltage and current phasors

Figure 5.3   Serial replacement diagram of the insulation system.

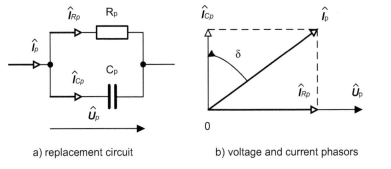

a) replacement circuit   b) voltage and current phasors

Figure 5.4   Parallel replacement scheme of the insulation system.

If we start from the parallel substitution scheme of the insulation system in Figure 5.4, where R_p is the *parallel insulation loss resistance* and C_p is the *parallel insulation capacity*, then for the *admittance* $\hat{Y}_p$, the impedance $\hat{Z}_p$ and the loss factor *tgd* of the insulation system

$$\hat{Y}_p = \frac{1}{\hat{Z}_p} = \frac{\hat{I}_p}{\hat{U}_p} = \frac{1}{R_p} + j \cdot \omega \cdot C_p, \qquad tg\delta = \frac{\hat{I}_{Rp}}{\hat{I}_{Cp}} = \frac{1}{\omega \cdot R_p \cdot C_p}. \quad (5.5a, b)$$

Based on the above equations, it is possible to determine the recalculations of the parameters of the elements of serial and parallel spare circuits

$$C_s = \left(1 + (tg\delta)^2\right)C_p, \qquad R_s = \frac{1}{1 + (tg\delta)^2}R_p, \qquad (5.6a, b)$$

$$C_p = \frac{1}{1 + (tg\delta)^2}C_s, \qquad R_p = \left(1 + (tg\delta)^2\right)R_s. \qquad (5.7a, b)$$

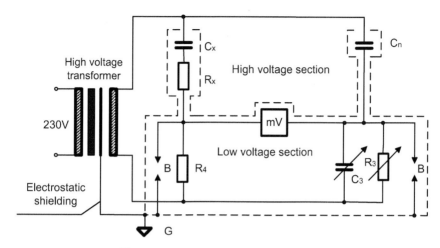

Figure 5.5 High voltage Schering bridge.

5.2.1 Measurement of capacity and loss factor by the HV Schering bridge

Either high-voltage *Schering bridges* or *high-voltage impedance meters* are used to measure the capacity and loss factor of HV insulation systems of electrical machines.

In terms of connection of insulation systems, we distinguish between the connection of the *Schering bridge* with the *ungrounded* and *grounded insulation system* of HV machines.

The *high-voltage Schering bridge*, designed to measure the capacity and loss factor of an ungrounded HV insulation system of an electric machine, is shown in Figure 5.5.

The *high-voltage part* of the bridge consists of the measured insulation system of the machine, characterized by its series capacity C_x and series loss resistance R_x, and a normal capacitor C_n with a negligible loss factor.

The *low-voltage part* of the bridge consists of a resistor R_4 and the resistance and capacitance decades C_3 and R_3, which balance the bridge, so that there is a minimum voltage on its measuring diagonal, which is indicated by the *zero indicator* NI. In simpler cases, the AC millivoltmeter mV is used as the zero indicator. In more demanding applications, *selective voltmeters* with a frequency selectivity equal to the frequency of the excitation voltage of the bridge (50 Hz) are used to indicate the voltage on the measuring diagonal.

In this way, it is possible to effectively suppress the influence of higher harmonic components, which arise from the nonlinearity of the conversion characteristics of the HV transformer.

Spark gap B is connected in the *measuring diagonal* nodes, serving to protect the low-voltage part of the bridge and the operator in the event of a breakdown of the bridge insulation system. In addition, the low-voltage part of the bridge is electrostatically shielded due to the effect of suppression of electromagnetic interference on the measurement accuracy. This grounded shield is connected to the electrostatic shield between the primary and secondary windings of the HV power transformer.

Assuming that the loss factor of the normal capacitor C_n is significantly lower in comparison with the loss factor of the measured insulation system of the machine, the capacity and loss resistance of the measured insulation system of the HV machine are

$$C_x = C_n \frac{R_3}{R_4}, \qquad tg\delta_x = \omega \cdot R_3 \cdot C_3. \qquad (5.8a, b)$$

In practice, it is necessary to correct the measured value of capacity by the capacity of the HV supply cable when the insulation system is not connected. The capacity of the HV supply cable is in the range of tens of pF/m of its length.

The *high-voltage Schering bridge*, designed to measure the capacity and loss factor of a *grounded HV insulation system* of an electrical machine, has one of the measuring diagonal nodes grounded (Figure 5.6).

The measurement of the capacity and loss factor of the insulation system is carried out in two stages. First, the measured insulation system is disconnected and the capacitance C_k and the loss factor $tg\delta_k$ of the supply cable are measured, including parasitic capacitances and bridge leads. After connecting the measured insulation system, the total capacity C_m and the loss factor $tg\delta_m$ are measured. It then applies to the capacity and loss factor of the insulation system itself

$$C_x = C_m - C_k, \qquad tg\delta_x = \frac{C_m \cdot tg\delta_m - C_k \cdot tg\delta_k}{C_x}. \qquad (5.9a, b)$$

Instead of normal capacitors, these bridges use special *ceramic* or *vacuum capacitors* up to 15 kV with nominal capacitances in the range of hundreds of pF to μF units with a negligible loss factor, reaching 10^{-5} to 10^{-6} at a frequency of 50 Hz. SF_6 (cheese fluoride) gas dielectric capacitors are used for higher voltages.

In practice, automatically balanced bridges are used for more convenient and faster measurement of capacitances and loss factors of insulation

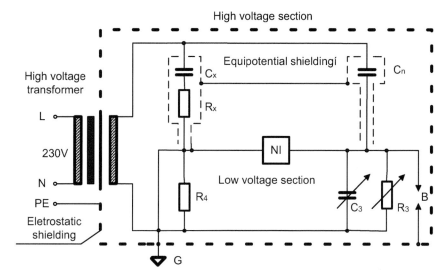

Figure 5.6   Shielded high voltage Schering bridge.

systems of electrical machines, in which a computer controls the above-mentioned measurement algorithm.

5.2.2 Measurement of capacitance and loss factor of HV impedance meters

High-voltage impedance meters consist of an HV source of alternating voltage, which supplies through current transformers the measured impedance $\widehat{\mathbf{Z}}_x$ of the machine insulation system and the normal impedance $\widehat{\mathbf{Z}}_n$, formed by a lossless HV capacitor C_n (Figure 5.7).

   To connect both impedances to the impedance meter, *equipotential shielding* is used to suppress the influence of capacitance and leakage resistance of the supply cables, in which the inner shielding sheath of the HV *triaxial cable* is at a potential equal to the measured voltage and the outer sheath of the cable that is connected to the *signal ground* (SG) is grounded. Because the voltage difference between the middle conductor and the inner sheath of the cable is negligible (given by the input residual voltage of the operational amplifier), the influence of leakage resistance and cable capacity on the measurement accuracy is practically suppressed.

   Impedance ratio measurement is often used to measure the parameters of insulation systems.

   When using series replacement circuits of measured and normal impedance, the division ratio $\hat{p}_s$ of their impedances is determined by series

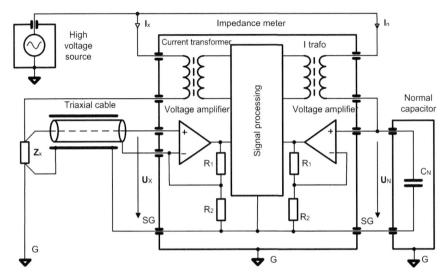

Figure 5.7   High voltage impedance meter.

replacement capacitances C_{xs} and C_{ns} and series loss resistances R_{xs} and R_{ns} of these impedances.

$$\hat{p}_s = \frac{\hat{Z}_{xs}}{\hat{Z}_{ns}} = \frac{C_{ns}}{C_{xs}} \frac{1 + j \cdot \omega \cdot R_{xs} \cdot C_{xs}}{1 + j \cdot \omega \cdot R_{ns} \cdot C_{ns}}. \tag{5.10}$$

If a normal capacitor C_n with a negligible series loss resistance R_{ns} is used in place of the normal impedance, then the capacitance and loss resistance of the measured impedance apply to

$$C_{xs} = \frac{C_n}{Re(\hat{p}_s)}, \qquad R_{xs} = \frac{Im(\hat{p}_s)}{\omega \cdot C_{xs}}. \tag{5.11a, b}$$

When using parallel spare connections of measured and normal impedance, the division ratio $\hat{p}_p$ of their admittances is determined by parallel, spare capacitances C_{xp} and C_{np} and parallel loss resistances R_{xp} and R_{np} of these impedances.

$$\hat{p}_p = \frac{\hat{Y}_{xp}}{\hat{Y}_{np}} = \frac{R_{np}}{R_{xp}} \frac{1 + j\omega \cdot R_{xp} \cdot C_{xp}}{1 + j \cdot \omega \cdot R_{np} \cdot C_{np}} \tag{5.12}$$

If the normal impedance is a capacitor C_n with an extremely high parallel loss resistance R_{np}, i.e., with a negligible loss factor, then for the capacity and loss resistance of the measured impedance

$$C_{xp} = C_n \cdot Re(\hat{p}_p), \qquad R_{xp} = \frac{1}{\omega \cdot C_n \cdot Im(\hat{p}_p)}. \tag{5.13a, b}$$

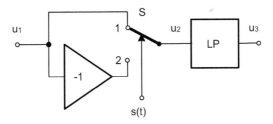

Figure 5.8 Switched detector.

Instead of the normal impedance $\widehat{Z}_n$, high-quality ceramic or vacuum HV capacitors with negligible losses are also used for these meters, which either are built into the meter or can be externally connected to it, preferably again with a *triaxial cable*.

Because it applies to measured and normal impedance

$$\hat{Z}_X = \frac{\hat{U}_X}{\hat{I}_x}, \qquad \hat{Z}_n = \frac{\hat{U}_N}{\hat{I}_N}, \qquad (5.14a, b)$$

where $\hat{U}_X$ and $\hat{U}_N$ are voltage phasors at impedances and and are phasors of currents flowing through these impedances

$$\hat{Z}_X = \frac{\hat{U}_X}{\hat{I}_x} \frac{\hat{I}_N}{\hat{U}_N} \hat{Z}_N. \qquad (5.15)$$

Awitched *detectors* are used to measure capacity and loss factor with direct *digital signal processing*.

Switched synchronous detectors are used in older impedance meters to measure real and imaginary components of impedance and loss factor. Modern impedance meters use *digital signal processing methods*.

Switched synchronous detectors are, in principle, formed by an amplifier with switchable gain ±1 and a *low-pass (LP) filter* (Figure 5.8).

The control signal $s(t)$ of the switch is derived from the HV AC voltage source. If the control signal has a positive polarity, i.e., $s(t) > 0$, the switch is in position 1 and the amplifier transmission is +1. If the control signal has a negative polarity, i.e., $s(t) < 0$, the switch is in position 2 and the amplifier transmission is −1. Because the time course of the control signal $s(t)$ with amplitude ±1 can be expressed by the time evolution

$$s(t) = \frac{4}{\pi} \sum_{n=0}^{\infty} \frac{1}{2 \cdot n + 1} \sin(2 \cdot n + 1) \omega t, \qquad (5.16)$$

the output voltage of the switched detector is determined by the product of the input voltage of the amplifier

$$u_1(t) = U_{1m} \cdot \sin\omega t \left(\omega \cdot t + \varphi\right) \tag{5.17}$$

with control signal $s(t) = \text{sign}[u_1(t)]$

$$u_3(t) = u_1(t) \cdot s(t) = U_{1m} \sin(\omega \cdot t + \varphi) \frac{4}{\pi} \sum_{n=0}^{\infty} \frac{1}{2 \cdot n + 1} \sin(2n + 1)\omega t. \tag{5.18}$$

It is clear from the above equation that the signal at the output of the detector contains, in addition to the *basic spectral component* with a circular frequency ω, also *higher odd harmonic components*, the amplitudes of which are inversely proportional to the values of their order. A low-pass filter with a cutoff circular frequency $\omega_m \leq 2\omega$ filters out these components.

At zero phase shift $\varphi = 0$, the mean value of the detector output voltage determines the *real component* of the voltage phasor $\hat{U}$ with respect to the signal $s(t)$. At the phase shift $\varphi = \pi/2$, the mean value of the detector output voltage determines the *imaginary component* of the voltage phasor with respect to the signal $s(t)$

$$\text{Re}(\hat{U}_1) = \frac{2}{\pi} U_{2m} \cdot \cos\varphi, \qquad \text{Im}(\hat{U}_1) = \frac{2}{\pi} U_{2m} \cdot \sin\varphi. \tag{5.19a, b}$$

Impedance meters with *digital signal processing* digitize voltage and current phasors A/D converters with a resolution of 16–24 bits, which operate with sampling frequencies up to several tens of kSps (*samples per second*). The signals thus digitized are digitally multiplied by the digitized harmonic signal of the excitation HV source.

If the non-phase-shifted digitized harmonic excitation HV voltage multiplies the digitized voltages and currents of the measured and normal impedances, then the mean value of the products of these voltages and currents with the excitation signal determines their *real components*. If these voltages and currents are multiplied by a harmonic excitation signal *shifted* by 90°, then the mean values of the products of the voltages and currents with the excitation signal are determined by the *imaginary components* of the measured and normal impedance.

A typical HV impedance meter is, for example, the TETTEX 2818 HV meter, which allows measuring the properties of insulation systems of MV machines up to 50 kV in the frequency range of 15 Hz to 1 kHz, capacitances in the range of 1 pF to 10 μF with an accuracy of 0.5%, and loss factor in the range of 10^{-5} to 1 with an accuracy of up to 2% at the frequency of the excitation HV voltage 50 Hz (Figure 5.9).

Figure 5.9 Tettex 2818 MV impedance meter.

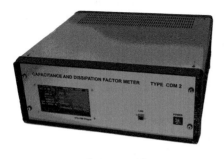

a) rear panel

b) rear panel

Figure 5.10 HV impedance meter developed at CTU FEE in Prague.

A similar device was developed at the Department of Measurement, Faculty of Electrical Engineering, and Czech Technical University in Prague within the solution of the TAČR project (Figure 5.10).

Two current transformers wound on toroidal cores are used to measure the currents flowing through the measured capacitances. The taps on their primary windings allow to measure currents in the ranges 16 mA, 160 mA, 1.6 A, and 16 A. A ferromagnetic cover shields the current transformers.

Both impedance meters are able to measure *ungrounded objects* in UST (*unground specimen test*) mode, to measure *grounded objects* in GST (*grounded specimen test*) mode, and to measure grounded objects with *active shielding* in GSTQ (*grounded specimen test quarding*) mode.

A *MEIDEN internal HV vacuum capacitor* with a nominal capacity of 100 pF and a nominal voltage of 21 kV with a loss factor of less than 10^{-6} is used in place of the impedance reference standard. An external capacitive normal capacitor can be used for ratio measurement of impedance and loss factor, which can be connected via the EXT C terminal (Figure 5.11).

Specification		
Peak Test Voltage	35kV,peak (at 50,60 Hz/ 1 minute)	
RF Working Voltage	21kV,peak (at 13.56 MHz)	
Capacitance	100 pF	
Maximum Current (Forced air)	126 A,rms (at 13.56 MHz)	
Total Length	73 mm	
Outer Diameter	φ48 mm	
Mounting Length	73 mm	
Weight	0.5 kg	

Figure 5.11   Vacuum condenser and its parameters.

Two 24-bit delta sigma A/D converters with sampling frequencies of 250 kSps are used in the meter to digitize voltages and currents. FPGAs are used to digitally process signals from these converters. A non-contact matrix LCD display is used to set the operating mode of the device, its ranges, and to display the measurement results. An optically isolated ETHERNET 1GBE interface is used to communicate the impedance meter with the computer. The device allows to measure capacitances in the range of 100 pF to 4 μF with an accuracy of 0.1% and a loss factor in the range of 10^{-5} to 1 with an accuracy of 1% at an excitation voltage of 50 Hz.

5.2.3 Measurement of frequency dependence of capacity and loss factor

This method, also known as the FDS (*frequency domain spectroscopy*) method, is designed to measure the frequency dependence of the capacity and loss factor of the insulation systems of the tested machines, most often HV transformers. The method is based on the method of measuring the impedance of HV insulation systems of machines by digital method, with the difference that in this method, the frequency range of the excitation sine signal is in the range of several tens of 10^{-6} Hz to tens of kHz and its effective value reaches tens of volts.

The meter consists of a generator of excitation sinusoidal voltage, consisting of a D/A converter and a memory in which this waveform is stored, power amplifiers VZ, and two current converters I/U_1 and I/U_2, whose voltage outputs are digitized by A/D converters ADC1 and ADC2 (Figure 5.12).

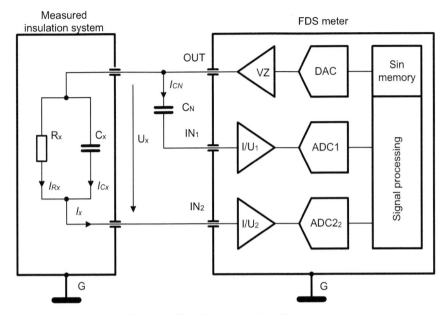

Figure 5.12 FDS meter block diagram.

The capacity of the insulation system is determined by the ratio of the RMS value of the imaginary component of the current phasor I_{Cx} of the measured insulation system to the RMS value of the current I_{CN} flowing through the normal capacitor C_N

$$C_x = \frac{I_{Cx}}{2 \cdot \pi \cdot f \cdot U_x} = \frac{I_{Cx}}{I_{CN}} C_N, \tag{5.20}$$

since the voltage U_x at the terminals of the insulation system is identical to the voltage at the normal capacitor

$$U_N = \frac{I_{CN}}{2 \cdot \pi \cdot f \cdot C_N}. \tag{5.21}$$

The current inputs of the current amplifiers I/U_1 and I/U_2 are at the signal ground potential of the SG device, which, however, does not have to be connected to the ground of the measured insulation system G.

The *loss factor* of the insulation system is determined by the ratio of the RMS value of the *real component* of the current phasor I_{Rx} to the RMS value of the *imaginary component* of the current phasor I_{Cx}

$$tg\delta = \frac{I_{Rx}}{I_{Cx}}. \tag{5.22}$$

Figure 5.13   Megger Frax 101 frequency and loss factor frequency meter.

Figure 5.13 shows a typical frequency and loss factor frequency meter FRAX101 from Megger, allowing to measure capacitance in the range of 10 pF to 100 µF with an accuracy of 0.5% + 1 pF and loss factor in the range from 0 to 1 with an accuracy of 1% in the frequency range of 1 mHz to 100 Hz. The device is able to generate test voltages up to 200 V with a maximum current of 50 mA in the frequency range of 50 µHz to 5 kHz.

5.3 Measurement of Dielectric Absorption of Insulation Systems

The *dielectric absorption* of an insulator is caused by a delay in the polarization of the dielectric as the polarity of its electric field changes. Because of this delay, there is no change in the charge of the insulator immediately after the voltage on the electrodes of the insulator, and a residual absorption charge remains in the insulator, which is manifested by the *residual voltage* on its electrodes.

The replacement scheme of the insulation system, taking into account the dielectric absorption, contains, in addition to its insulation resistance R_{iz}, capacity C_s and loss resistance R_s, also its absorption capacity C_a and absorption resistance R_a (Figure 5.14).

Since $R_s \cdot C_s \gg R_a \cdot C_a$ holds for the time constants, after connecting the system to the step change of voltage U_s, the time course of the charging current $i(t)$ is approximately exponential with the time constant $R_s \cdot C_s$ and its maximum value U_s/R_a is at $t = 0$ limited by absorption resistance R_a. The exponential course of the charge $q(t)$, accumulated in the insulation system, also corresponds to this exponential course of the current. At steady state, the DC insulation current is determined by the ratio

$$I_{iz} = \frac{U_s}{R_{iz}}. \tag{5.23}$$

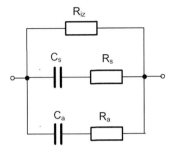

Figure 5.14   Replacement diagram of insulation system with dielectric absorption.

Figure 5.15 shows the time courses of the test voltage, current, and charge during charging and discharging of the insulation system.

The *dielectric absorption* of the insulation system is most often expressed by the following parameters.

Polarization index (PI) is defined by the ratio of charging currents after 15 and 60 s after the supply voltage is connected (Figure 5.16).

$$PI\left(\frac{15}{60}\right) = \frac{i_d(15)}{i_d(60)}. \tag{5.24}$$

In addition to this 1-minute *polarization index*, a *10-minute polarization index* is determined when measuring quality insulation, defined by the ratio of charging currents after 60 and 600 s from the supply voltage connection.

$$PI\left(\frac{60}{600}\right) = \frac{i_d(60)}{i_d(600)}. \tag{5.25}$$

The DAR (*dielectric absorption ratio*) is defined by the ratio of the charging currents after 30 and 60 s after the supply voltage is connected. In some cases, the dielectric discharge (DD) factor is also given, which, unlike the polarization indices measured when the dielectric is charged, is measured during its discharge. Therefore, it is not necessary to use a stable HV voltage source for this measurement. The *absorption charge* is defined by the difference of the areas of the time courses of its charging and discharging currents, excluding the areas corresponding to the insulation current I_{iz}.

Figure 5.17 shows a typical operating meter of insulation current, insulation resistance, and polarization S1-568 from MEGGER. The device allows to measure insulation currents up to 5 mA and insulation resistance in the range of 10 kΩ to 35 TΩ in the voltage range of 5–15 kV at 25-V step changes. The device also allows to measure insulation capacitances in the range of 100 nF to 25 μF and polarization indices in adjustable times up to 600 s.

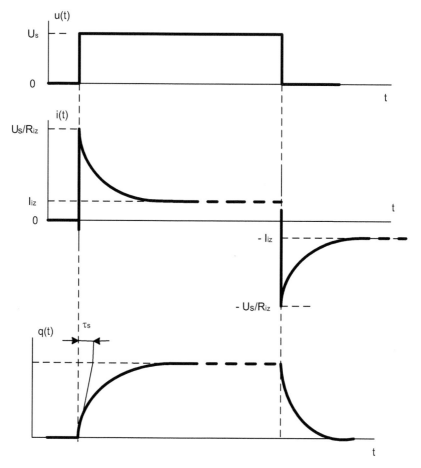

Figure 5.15   Time courses of voltage, current and charge during charging and discharging of the insulation system.

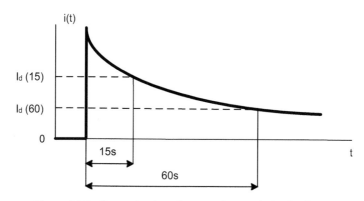

Figure 5.16   Determination of a one-minute polarization index.

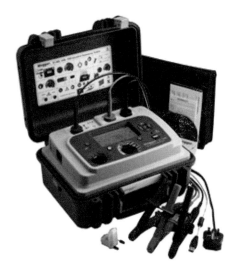

Figure 5.17 Leakage current, insulation resistance and polarization operating meter MEGER S1-568.

References

[1] Trigeassou, J.C. *Electrical Machines Diagnosis*. Wiley ISTE 2011. ISBN: 978-1-848-21263-3-

[2] Vedral, J., Svatoš, J. *Analog signal processing and digitization in measurement*. CTU Publishing House 2020. ISBN: 978-80-01-06717-8.

[3] Vedral, J. *Digital measuring instruments of electrical quantities*. Publishing House 2021. ISBN: 978-80-01-06860-1.

[4] Clyde, F. C. *Electronic Instrumemnt Handbook*, 7. Edition, Mc Graw-Hill 1994, ISBN: 0-07-012616-X.

[5] Northrop, R.B. *Introduction to Instrumentation and Measurements*, CRC 2005. ISBN: 0-8493-3773-9.

[6] Placko, D.: *Fundamentals of Instrumentation and Measurement*, ISTE 2007, ISBN: 1-905209-39-8.

[7] *Low Level Measurements Handbook* - 7th Edition Precision DC Current, Voltage, and Resistance Measurements, Keithley 2018.

[8] *High Voltage Schering Bridge*. https://www.eeeguide.com/high-voltage-schering-bridge/

[9] Lemke Diagnostic GMBH. *User manual*: LDV - 5: *Capacitance and loss factor measuring*. Volkersdorf, 2003.

[10] Haefely Hipotronics. *Precision Oil and Solid Dielectric Analyzer*, Basel 2016. https://update.haefely.com/ct2830/_private/MA_2830_2831_4843477_V2.0.pdf.

[11] Mittal, M., Sridhar, M., Tamrakar, R., Agarwal, S., Rajput, S. *Innovative system for precision measurement of high voltage capacitance & tan δ.* International Journals of Multi-Dimensional Research. Haryana 2014. http://ijmr.net.in/download.php?filename=PuXXl2Wc51dGP3N. pdf&new=IJPA2462Feb5.pdf.

[12] Tereň, O., Tomlain, J., Sedláček, R., Vedral, J. *Capacitance and dissipation factor meter for high-voltage electrical machines,* XXI IMEKO World Congress, CTU in Prague 2015, ISBN 978-80-01-05793-3.

[13] Tereň, O., Tomlain, J., Sedláček, R., Vedral, J. A Hardware Platform for Frequency Domain Spectroscopy and Frequency Response Analysis, Acta Polytechnica Hungarica, Vol.14, No. 8, 2017, pp. 47 – 63.

6

Partial Discharges in Insulating Systems of HV Electric Machines

---

Partial discharges are a special type of electrical discharges according to IEC 60 270, in which there is a partial violation of part of the dielectric insulation. With these discharges, the remaining part of the dielectric still has sufficient electrical strength to allow further trouble-free condition of the electrical machines.

6.1 Formation and Types of Partial Discharges

Partial discharges are caused by partial punctures of dielectrics due to their different electrical strengths. The most common source of partial discharges are gaseous, liquid, or solid dielectrics with local gas cores. The *ignition voltages* at which partial discharges occur depend on not only the electrical strength and permittivity of the gases in the cavities and insulator but also on the size and shape of the core. The source of partial discharges is also inhomogeneities arising from the contact of the insulator with the metal electrode. In terms of the place of occurrence, we therefore distinguish *external, surface (sliding), internal,* and *tree partial discharges* (Figure 6.1).

Figure 6.1a shows the formation of *external discharges,* e.g., resulting from the tip-plate arrangement of electrodes. These so-called *corona discharges* are formed in gases at sharp points in the electric field. They usually occur on the high voltage side but can also appear on sharp edges with ground potential. Although the distance between the electrodes can be large, a high concentration of the field at the tip can cause partial puncture. Therefore, to prevent the occurrence of these discharges, it is necessary to choose such design solutions that do not

Figure 6.1b shows the formation of a surface discharge formed at the interface of the electrode with the gas. These discharges are formed under

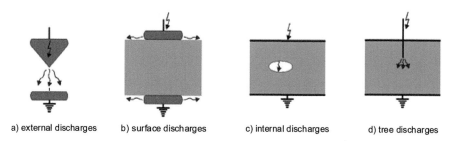

a) external discharges   b) surface discharges   c) internal discharges   d) tree discharges

Figure 6.1   Typical electrode arrangements for partial discharges.

electrical stress; I acts parallel to the dielectric surface. Surface discharges are generated mainly on transformer bushings, cable terminals and surface contacts. In addition, they affect the electric field beyond the place of their origin; so they can spread beyond this place.

Figure 6.1c shows the formation of *internal partial discharges*. Gas cores are located in the insulating material itself or just below the electrode during the so-called delamination (separation of layers) of individual materials with each other. *Delamination* occurs either during the production of insulation or during operation under the action of high temperature and pressure on individual layers of insulation.

Figure 6.1d shows the formation of *treeing discharges*, the occurrence of which is conditioned by defects in insulating materials. These trees often grow to larger sizes and can also be filled with gaseous cavities. Partial discharges occur in these cavities, and, thus, combined types of partial discharges are formed, which are unstable in time and often lead to complete destruction of the insulation properties of the insulation.

Table 6.1 shows the basic types of partial discharges in their elliptical display according to IEC 60 270.

The elliptical representation of partial discharges consists in displaying their course, including the test voltage, using *Lissajous patterns*; see Section 7.1.3.1 in Chapter 7.

Figure 6.2 shows the time courses of typical partial discharges with a color expression of the frequency of occurrence of their charges, while the warmer colors correspond to the higher frequency of occurrence of charges.

Figure 6.3 shows the distribution of electric fields of two identical plate insulators before and after the discharge, between which two spherical insulators with a permittivity smaller than that of the insulators surrounding them are placed.

Table 6.1 Basic types of partial discharges in their elliptical display according to IEC 60 270.

type	Oscilloscopic image	Description	Discharge type
A		Pulses of the same size in one half-period, symmetrically around the voltage maximum. With increasing voltage, the number of pulses increases, but their size does not change. In the opposite half-period, impulses do not occur.	Tip-plate (rod-plane) arrangement in gases. If the impulses appear in a negative half-period, the tip (rod) is at a high potential. If the impulses appear in positive half-period, is the tip (rod) on the ground potential.
B		Tip-plate (rod-plane) arrangement in gases. If the impulses appear in a negative half-period, the tip (rod) is at a high potential. If the impulses appear in positive half-period, is the tip (rod) on the ground potential.	Tip-plate arrangement in liquid dielectrics. If the larger impulses are in a positive half-period, then the spike is at a high potential. If the larger impulses are in a positive half-period, then the spike is at a high potential.
C		The impulses between the passes of zero voltage and the vertices in both half-periods have approximately the same amplitudes.	Cores in solid or liquid dielectric. Touch of insulated wires. Surface discharges on the surface of the insulator without galvanic connection. Ungrounded metal parts of the insulation system.
D		Impulses between passages through zero voltage and peaks in both half-periods. The impulses in one of the half-periods are greater than the impulses in the other half-period of voltage.	Air cores in solid and liquid insulators. If large pulses are in a positive half-period of voltage, the discharges are at a high potential. If large pulses are in a negative half-period of voltage, the discharges are on the ground potential.
E		Pulses symmetrically distributed around zero voltage passages.	Defective contact between metal parts or between semiconducting layers of the insulation system.

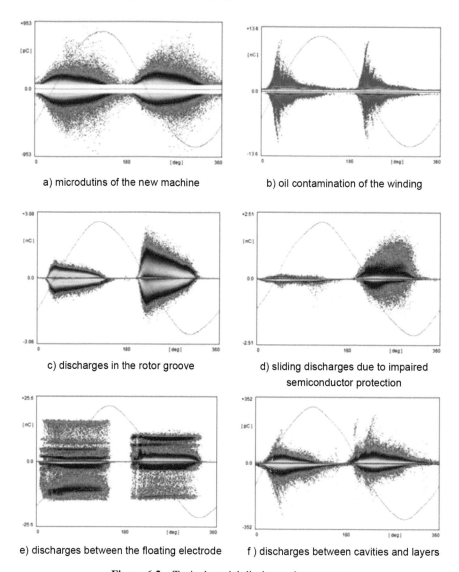

a) microdutins of the new machine

b) oil contamination of the winding

c) discharges in the rotor groove

d) sliding discharges due to impaired
semiconductor protection

e) discharges between the floating electrode

f) discharges between cavities and layers

Figure 6.2   Typical partial discharge time courses.

6.2 Partial Discharge Parameters

The basic electrical parameters of partial discharges, defined by EN 60270, include for *individual discharges* their apparent charge, the time moment of their occurrence, referred to as the phase angle, and incendiary and extinguishing voltage. The *sequence of partial discharges* is defined by the

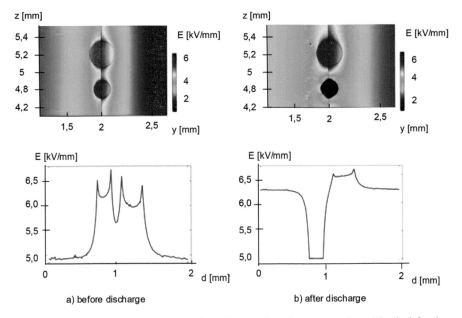

Figure 6.3 Distribution of the electric field strength at the contact of two identical fixed.

frequency of their occurrence, the sum charge, the mean current, the mean quadratic sum, and the power of the discharges.

Apparent charge q_z is defined as a charge that produces the same reading as the actual partial discharge pulse in the insulation system but which cannot be detected because it is inside the insulation system and is therefore inaccessible. The apparent charge reaches units up to hundreds of nC.

Repetition rate n is the ratio between the total number of current pulses m produced by the discharge activity over a certain time interval and the duration of that time interval T_r

$$n = \frac{m}{T_r} \left[s^{-1} \right].$$ (6.1)

In practice, when determining the frequency of impulses, only those impulses that are above a certain level or are located in the prescribed amplitude interval are considered.

Average discharge current I is the current given by the sum of the absolute values of the individual charges of the partial q_i discharges during their duration T

$$I = \frac{1}{T} \sum_{i=1}^{m} |q_i|,$$ (6.2)

where *m* is the number of discharges in the time interval *T*. Mean current of partial discharges reaches tens of microamperes to units milliamperes.

If all discharges have the same apparent charge *q*, the mean value of the partial discharge current is

$$I = nq \ [A],$$ (6.3)

where *n* is the frequency of their occurrence.

Cumulative charge Q gives the sum of the absolute values of the individual sizes of the apparent charges over a certain time interval

$$Q = \sum_{i=1}^{m} |q_i| [C].$$ (6.4)

The sum charge is usually related to one or more periods of test voltage.

The *phase shift* φ_- of the *i*th occurrence of partial discharges φ_i is defined by the following equations:

$$\varphi_i = \frac{t_i}{T_p} [\text{rad}], \qquad \varphi_i = 360 \frac{t_i}{T_p} [°].$$ (6.5a b)

where t_i is the time between the previous positive pass of the test voltage zero and the moment of occurrence of the partial discharge and T_p is the period of the test alternating voltage.

The *partial discharge power P* indicates the mean value of the power of charge pulses during a defined time interval *T*, which is usually the time of the excitation voltage period

$$P = \frac{1}{T} \sum_{i=1}^{m} q_i u_i,$$ (6.6)

where u_1, u_2, ..., u_i are the instantaneous values of the test voltage at the moments of occurrence of individual discharges, and *m* is the number of discharges in the time interval *T*.

This time interval is usually the time of the period of excitation of voltage. The power of partial discharges ranges from tens to hundreds of megawatts.

The *mean square sum* of charges is a derived quantity that is given by the sum of the quadrates of the individual charges of the apparent q_i discharges during the selected time interval *T*, which is usually the period of the excitation voltage

$$Q_{mss} = \sqrt{\frac{1}{k} \sum_{i=1}^{k} q_i^2}.$$ (6.7)

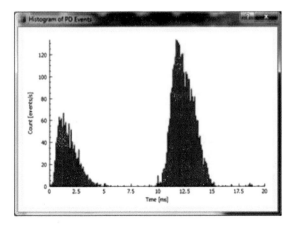

Figure 6.4 Frequency of occurrence of partial discharges during one period of test voltage.

If all charges are of the same magnitude q, then their *mean quadratic sum* is

$$Q_{mgs} = nq, \tag{6.8}$$

where n is the frequency of occurrence of partial discharges.

In addition, the incendiary and extinguishing voltages of the discharges are defined.

Ignition voltage U_{ign} is such a voltage from which repetitive partial discharges are observed when increasing the test voltage.

Extinguishing voltage U_{ext} is a voltage at which repetitive partial discharges cease to occur when the test voltage is reduced.

For a more detailed assessment of the properties of partial discharges, the determination of obliquity, sharpness, number of vertices, charge, and phase asymmetry from the frequencies of their occurrences in a given time interval is used.

Figure 6.4 shows examples of partial discharge frequencies during a time interval of 20 ms.

The *skewness* (SK) distribution of the frequency of occurrence of partial discharges describes its asymmetry to the normal (*Gaussian*) distribution

$$SK = \frac{\sum_{i=1}^{k} p_i (x_i - \mu)^3}{\sigma^3}, \tag{6.9}$$

where p_i is the *probability of occurrence* of the amplitude of discharge x_i in the ith interval, μ is the *mean value* of the frequency of occurrence of partial discharges

$$\mu = \sum_{i=1}^{k} x_i p_i, \tag{6.10}$$

and σ is the *mean quadratic deviation* of this frequency for which

$$\sigma = \sqrt{p_i \left(x_i \mu\right)^2}.$$ (6.11)

If the shape of the frequency of occurrence of amplitudes of partial discharges is symmetrical, then the obliquity of this distribution is zero. If this is not the case, then when the left part of the layout is deflected to the left, the obliquity of the layout is positive, and when the right part of the layout is deflected to the right, the obliquity of the layout is negative.

The *kurtosis* (KU) characterizes the deviations in the shape of the frequency of occurrence of partial discharges relative to the normal distribution

$$KU = \frac{\sum_{i=1}^{k} p_i \left(x_i - \mu\right)^4}{\sigma^4}.$$ (6.12)

If the frequency of occurrence of partial discharges is identical to the normal distribution, then the sharpness of this distribution is zero. If this layout has a flatter course than the normal distribution, then its sharpness is negative, and vice versa.

Number of peaks (NP) is used to distinguish character frequencies with one or more vertices. The peak frequency of occurrence of random discharges is determined by successive changes in polarities of partial discharge frequencies

$$\frac{dy_{i-1}}{dx_{i-1}} > 0, \qquad \frac{dy_{i+1}}{dx_{i+1}} < 0.$$ (6.13a, b)

Charge asymmetry (CHA) determines the proportion of mean frequencies of occurrence of partial discharges in the positive and negative half-waves of the AC test voltage

$$CHA = \frac{Q_+ / n_+}{Q_- / n_-},$$ (6.14)

where Q_+ and Q_- are the sum charges in the positive and negative half-waves of the test voltage, and n_+ and n_- are the numbers of discharges in these half-waves.

Phase asymmetry (PA) determines the phase fractions when the initial ignition voltages are generated in the positive and negative half waves of the AC test voltage

$$PA = \frac{\varphi_{i+}}{\varphi_{i-}}.$$ (6.15)

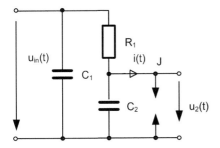

Figure 6.5 Model of external and surface partial discharge [5].

Knowledge of obliquity, sharpness, number of vertices, and charge and phase asymmetry of partial discharge frequencies allows their more detailed comparative analysis using sophisticated signal processing methods with elements of artificial intelligence.

6.3 Partial Discharge Models

To describe the properties of external, surface, and internal partial discharges, models are used, characterizing their properties when exciting by alternating voltage. The *Gemant–Philipps* or *Böning model* is used to describe internal partial discharges.

6.3.1 Model of external and surface partial discharges

The behavior of external and surface partial discharges is characterized by a model in which capacitor C_3 represents the capacity of the intact insulation system, capacitor C_1 the capacity of the surface failure of the insulator, resistor R_1 its resistance, and SS the spherical spark gap (Figure 6.5).

If the voltage on capacitor C_1 does not reach the *ignition voltage* of the insulation U_{ign}, then for the voltage on the capacitor C_1

$$u_{20}(t) = \frac{u_1(t)}{1 + j\omega R_1 C_2}. \tag{6.16}$$

If the voltage on the capacitor C_1 reaches the ignition value of U_{zap}, then on the spherical spark gap, the voltage will jump and the voltage on it will jump to zero. Due to the reduction of the electrical strength of the gas in this type of partial discharge, the jumps are repeated even during the time during which the voltage $u_{20}(t)$ is less than the ignition voltage of the insulation.

Figure 6.6 shows the time courses of voltage and current in external and surface partial discharges.

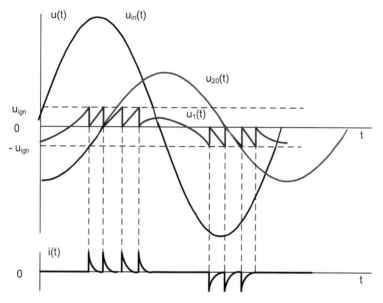

Figure 6.6 Time courses of voltage and current under external and surface partial discharge.

The *electric charges of surface discharges* are theoretically identical and equal to the charges

$$q = C_2 \cdot U_{ign},\qquad(6.17)$$

but the frequency of their occurrences increases with the growth of the test voltage. Similar properties have partial discharges at the included DC voltage.

The advantage of using DC voltage to measure partial discharges due to the use of alternating voltage is that with this method of excitation, there are significantly lower requirements for the power of the HV source.

6.3.2 Gemant–Philips model of internal partial discharges

This *three-capacity model* is the default model for analyzing the properties of internal partial discharges (Figure 6.7).

In this model, capacitor C_1 represents the capacitance of a separate core, capacitor $C_2 = (C_{21} \cdot C_{22})/(C_{21} + C_{22})$ represents the capacity of the residual part of the intact insulation, and the capacitor $C_3 = C_{31} + C_{32}$ represents the

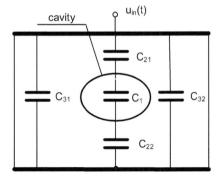

a) insulator with internal gas core

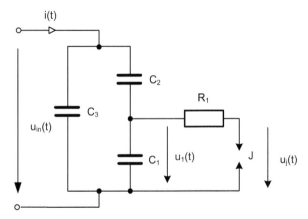

b) replacement scheme of insulator with inner cavity

Figure 6.7   Gemant-Phillips model dielectric.

capacity of the undamaged part of the dielectric. Resistor R_1 represents the resistance of the discharge channel. The time courses of voltages and currents in this model are shown in Figure 6.8.

If the voltage $u(t)$ does not reach its ignition value U_{ign}, then for the voltage on the core

$$u_{10}(t) = \frac{C_2}{C_1 + C_2} u(t). \qquad (6.18)$$

If the voltage on the core $u_1(t)$ exceeds the ignition value of the U_{ign} voltage, a discharge occurs in it. The voltage on the core drops to the *extinguishing voltage* U_{ex} and the discharge goes out. On all cavities of the dielectric, there is a voltage jump at different voltage values.

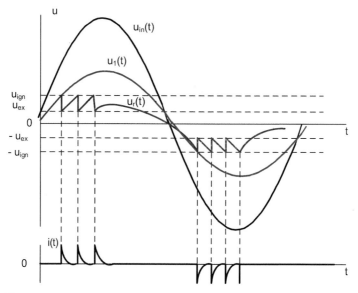

Figure 6.8   Time courses of voltage and current in the Gemant-Philips model of dielectric.

The duration of such a discharge is very short; it takes up to tens of nanoseconds in units of the order of magnitude, since during the discharge, there is a breakdown of neutral gas molecules in the cavity into positive and negative ions, which are the cause of the formation of currents determining the sizes of partial charges.

After the discharge is extinguished, the voltage on the cores increases again. If the voltage on them reaches the ignition voltage, other partial discharges are again formed. If the voltage on the cores is less than the ignition voltage, there are no further partial discharges. This phenomenon can be repeated several times during both positive and negative half-periods of exciting alternating voltage. Assuming that $C_3 \gg C_1 \gg C_2$, the *actual charge* is a partial discharge charge

$$q_s = (C_1 + C_2) \cdot (U_{ign} - U_{ex}).$$

(6.19)

However, this charge is directly immeasurable, since the discharges take place inside the insulator, and, in addition, we do not know the magnitudes of the capacities C_1, C_2, and C_3. Therefore, when measuring partial discharges, the concept of the so-called *apparent charge* is introduced, for which

$$q_z = \frac{C_2}{C_2 + C_3}\left(U_{ign} - U_{ex}\right) = q_s \frac{C_2}{C_1 + C_2}. \qquad (6.20)$$

The *energy of a partial charge* is determined by the difference in energies in capacitors C_1 and C_2 before and after its formation. Since $U_{zap} >> U_{zh}$, it applies to the energy of the partial charge

$$W = q_s\left(U_{ign} - U_{ex}\right) \approx q_s \frac{C_1 + C_2}{C_1} U_{ign}. \qquad (6.21)$$

6.3.3 Bönings model of internal partial discharges

This *five-capacity model* is used to more accurately describe discharge activity. The model is based on the idea that not only the surface of the core is involved in the discharge process but also a certain part of the surrounding material surrounding the core. During its breakthrough, charges are sucked out not only from the walls of the cavity but also from the material around the cavity. Therefore, there is no voltage drop only on the cavity but also in its surroundings.

Unlike the Gemant–Philips model, Böning's model is extended by capacitors C_4 and C_5 and resistor R_2, where capacitance C_4 represents the capacity of the surface of the core, capacity C_5 represents the capacity of the remainder of the insulator, and resistor R_4 determines the resistance of the walls of the core and its vicinity. The replacement scheme and time courses of voltage and current in the *Bönings model* are shown in Figure 6.9, and time courses of voltages and currents in this model are shown in Figure 6.10.

If the voltage $u_1(t)$ on the spherical spark gap does not reach the ignition value U_{ign}, then the voltages $u_1(t)$ and $u_4(t)$ are identical and the current does not pass through the resistance R_4. If the voltage $u_1(t)$ reaches the ignition value, a partial discharge occurs and the voltage on the core drops to the $U_{ign} - U_d$ value, where U_d is the voltage on the core located inside the insulator. This voltage drop is greater than the *recovery voltage* U_{rec}, which in the previous Gemant–Philippoff model replaces the *extinguishing voltage* U_{ex}.

If the duration of the discharge is Δt, then for the voltage on capacitors C_1 and C_4 at the end of this period

$$U_1\left(\Delta t\right) = U_{ign}\exp\left(-\frac{\Delta t}{R_1 C_1}\right), \qquad U_4\left(\Delta t\right) = U_{ign}\exp\left(-\frac{\Delta t}{R_4 C_4}\right). \qquad (6.22a, b)$$

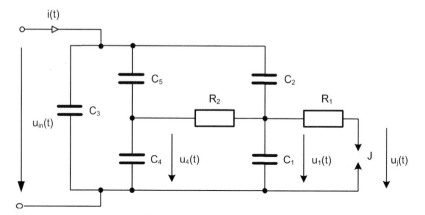

Figure 6.9 Replacement scheme of Bönings model.

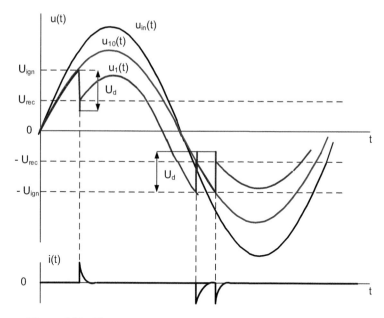

Figure 6.10 Time courses of voltage and current in the Böning model.

The ratio of recovered to incendiary voltage defines the ratio that is at $\Delta t \ll R_4 C_4$

$$p = \frac{U_{rec}}{U_{ign}} = \frac{1}{1 + C_1 / C_4} \exp\left(-\frac{\Delta t}{R_4 C_4}\right) \approx \frac{1}{1 + C_1 / C_4}. \qquad (6.23)$$

This ratio determines the *frequency of occurrence of partial discharges*

$$n = 4\frac{(U_{1m} / U_{ign}) - 1}{1 - p},$$ (6.24)

where U_{1m} is the amplitude of the voltage waveform $u_1(t)$.

According to this model, the *energy of the partial charge* is determined by the sum of the energy W_1 released during the discharge in the cavity and the energy ΔW_4, corresponding to the energy during the transfer of charge from capacitor C_4 to capacitor C_1 at a given value of *recovery voltage U_{rec}*. Assuming that $U_{zot} \approx 0$ and C_2, $C_5 \ll C_1$, C_4, applies to the partial discharge energy

$$W = \frac{1}{2}C_1(1 + p)U_{ign}^2.$$ (6.25)

where p ranges from 0.1 to 0.9 for this model. In the *Gemant–Philippoff model*, this ratio is zero.

Other models describing the issue of internal partial discharges are, e.g., *Kranz's model*, differing from *Böning's model* by replacing the resistance R_1 with a time- and voltage-dependent resistance. This model more accurately describes the course of stress on the cavity after the discharge is ignited and extinguished.

6.4 Effect of Partial Discharges on the Insulation Systems of HV Machines

Partial discharges by their effects cause changes in the physical and chemical properties of the insulator, resulting in *physical erosion* of the material, although the energy of the discharges is relatively low. These changes are usually irreversible, and damage to the insulation system is manifested by a decrease in its electrical strength.

Gaseous insulations, e.g., air and nitrogen, are very resistant to partial discharges, as their ionic conductivity is usually reversible.

Anorganic Insulating materials, e.g., porcelain, mica, and glass, have very good resistance to partial discharges. High temperature resistance also characterizes these materials. Their disadvantage is that they are fragile and therefore cannot be directly used in the insulation systems of HV electric machines. Therefore, they are used only as additives to insulating resins and textiles.

Organic materials such as polyethylene (PE), cross-linked polyethylene (XLPE), polytetrafluoroethylene (PTFE, Teflon), resin-based laminates, and materials using silicones as binders are very sensitive to partial discharges. Even with a short exposure to partial discharges, their insulating properties may be degraded or there may be irreversible breakthrough.

Electro-erosive, chemical, and thermal effects or their combinations manifest partial discharges in dielectrics.

6.4.1 Electro-erosive effects of partial discharges

They occur when an arc is formed in the cavity. Near the end of the discharge channel, the electric field has similar effects to the field near the sharp edge of the conductor. The field strength is very high and the field is very inhomogeneous. At a high intensity of an electric field, the concentration of this field can create an electric breakthrough. These punctures can create *carbon conductive pathways* in the cavities.

These conductive pathways from different cavities can merge to form *longer conductive chains*. The emerging and connecting channels have the shape of a so-called electric tree. Erosive effects also have electrons and ions, which bombard the walls of the cavity and cause its disruption. This increases the size of the core, which can take such values that it bridges the electrodes and a puncture occurs.

6.4.2 Chemical effects of partial discharges

By the chemical action of partial discharges, gases are formed in the gaseous, liquid, and solid states of insulators, which degrade their insulating properties.

In the case of air cavities, oxygen O_2 and nitrogen N are the most widely used. Oxygen decomposes into O_3 ozone under the action of discharges, which has significant oxidative effects. Nitrogen is formed by the action of discharges, so-called *active nitrogen*, which has much more erosive effects than ordinary nitrogen. Furthermore, *nitrogen oxides*, NO and NO_2, can be formed during discharges.

These nitrogenous oxides usually react with air humidity to form nitrogenous acids. Together with the action of high temperature, nitric acid (HNO_3) is formed, which causes intense erosion of the walls of the cavities.

Chemicals are formed initially only in the space of cavities. With the passage of time, however, these substances increase and these begin to spread to the areas surrounding the cores and thus disrupt not only the insulating strength of the material but also reduce its resistivity and thus increase its dielectric losses.

When discharges act in inhomogeneities of cross-linked polyethylene (XLPE), gaseous substances such as carbon monoxide (CO) and carbon dioxide (CO_2) are formed. In very small quantities, hydrogen (H_2) is also formed. If the atmosphere contains water and if carbon monoxide (CO) is present, drops of formic, acetic, and carboxylic acids are formed on the surface of the cross-linked polyethylene. These drops then become solid crystals under the action of discharges.

Similarly, liquids of a predominantly acidic character are formed on the surface of epoxies. These liquid products then penetrate into the insulators and increase the conductivity of the walls of their cavities by up to several orders of magnitude.

6.4.3 Thermal effects of partial discharges

The discharge channel is heated to a high temperature, which acts on the walls of the cavity. Under the action of this high temperature, the surface can be charred where this channel was formed. This increases the electrical conductivity of the surface of the core. As such, the discharge is associated with the release of a not-insignificant amount of thermal energy.

If the heat generated by the insulating material is not sufficiently well dissipated, the material may *overheat* and, in the worst case, there may be heat breakthrough. In the material of the insulator, both heat from discharges and heat generated by *dielectric losses* are generated. Together, they can help reduce the voltage required for thermal breakthrough. Increasing the temperature of the material has a positive effect on reducing the number of discharges, but on the other hand, the resulting discharges have a greater intensity.

Figures 6.11–6.13 show examples of accidents caused by HV transformers caused by partial discharges.

Figure 6.11 Destroyed encapsulated conductor outlet.

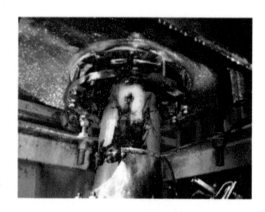

Figure 6.12   Destroyed block transformer 132 kVA.

Figure 6.13   Destroyed windig o HV transformer.

References

[1] Florkowski, M. *Partial Discharges in High-voltage Insulating.* Wydawnictwa AGH 2020, ISBN: 97-8836636-4752

[2] Hauschild, W., Lemke, E. High Voltage Test and Measuring Techniques. Springer Verlag. 2018. ISBN: 978-3-319-97459-0

[3] Bonning, W. *Luftgehalt und Luftspaltverteilung gesichterer Dielektrik,* part *I. Untersuchung der Entladung in einzelnen Luftspalten bei ausserem Wechselfeld.* Electrical Engineering, vol 48, No 1. Springe Verlag 1963. ISSN: 0948-7921

[4] Zalis, K. Evaluation of partial discharge measurememt. VII Symposium EUI 99. Zakopanie 21-23, Krakow 1999

[5] Germant, A., Philippoff, W. *Zeitung fur technische Physic 13 (1932),* p. 425.

[6] O'dwyer, J. J. *Theory of avalanche breakdown in solids with space charge distortion of the field.* Conference on Electrical Insulation &

Dielectric Phenomena - Annual Report 1969. IEEE, 1969, s. 137–140. ISBN 978-0-309-01764-0.

[7] Haefely Test AG. *Breaking the limit of power capacitor resonance frequency with help of PD pulse spectrum to check and setup PD measurement.* Haefely Hipotronics.hipotronics.com/document/breaking-the-limit-of-power-capacitor-resonance-frequency-with-help-of-pd-pulse-spectrum-to-check-and-setup-pd-measurement/

[8] Kebbabi, L. *Optical and electrical characterization of creeping discharges over solid/liquid interfaces under lightning impulse voltage.* IEEE Transactions on Dielectrics and Electrical Insulation Vol. 13, No. 3, June 2006

[9] König, D., Rao, Y.N. *Partial discharges in electrical power apparatus.* VDE-Verlag, Berlin 1993.

[10] IEC 60 270 *Partial discharge measurement.* IEC 2000

[11] ASTM D1868 *Standard Test Method for Detection and Measurement of Partial Discharge (Corona) Pulses in Evaluation of Insulation Systems*

[12] CIRGE Technical Brochure No. 182. *Partial Discharge Detection in Installed HV Extruded Cable Systems.* WG 21.26, 2001

[13] CIRGE Technical Brochure No. 366. *Guide for Electrical Partial Discharge Measurement in Compliance to IEC 60 270*, WG D1.33, 2008

7

Measurement and Detection of Partial Discharges

Partial discharges are accompanied by electromagnetic, acoustic, thermal, and chemical phenomena. In terms of the nature of measurement and detection of partial discharges, we distinguish *global* and *localization methods*. Global methods make it possible to measure the detailed properties of partial discharges in the insulation systems of HV electric machines. Localization methods allow you to determine the place of their formation.

In practice, when diagnosing insulation failures, the total discharge activity is first determined by global methods, and then the place of its origin is determined by localization methods.

Electrical and non-electrical methods are used to measure and detect partial discharges.

7.1 Electrical Methods Measurement and Detection of Partial Discharges

Electrical methods of measurement and detection of partial discharges include *galvanic methods*, measurement of partial discharges by *current transformers*, *inductive* and *capacitive probes*.

7.1.1 Properties of partial discharge current pulses

The principle of electrical methods of measuring partial discharges is the *measurement of their charges*. The electric charge of the current pulse is determined by its integration in the range of time moments t_1 and t_2

$$q = \int_{t_1}^{t_2} i(t)dt. \tag{7.1}$$

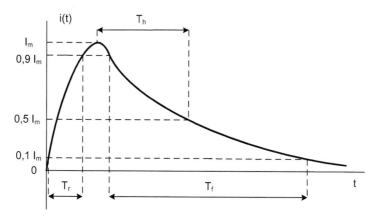

Figure 7.1   Time course of typical partial discharge current.

The *current pulses* of partial discharges are characterized by the *rise time T_r*, defined by the time it takes for the current to rise from 0.1 to 0.9 of its maximum value, and the *fall time T_f*, given the time it takes for the current to fall from 0.9 to 0.1 of its maximum value. Instead of the descent time, when assessing the time courses of partial discharges, the *half-rear time T_h* is defined, given by the time it takes for the amplitude of the partial discharge to fall to 1/2 of its maximum value (Figure 7.1).

In *solid dielectrics*, the start time of current impulses of partial discharges reaches several tens of nanoseconds and rear times up to tens of microseconds. In *liquid dielectrics*, the rise time of the current impulses of the ns units and the rear time of the µs units.

Based on the knowledge of the start times and descent times of the current pulses of partial discharges, it is possible to determine the *upper* and *lower limit frequencies* of current pulses

$$f_{mh} = \frac{0{,}35}{T_n}, \qquad f_{md} = \frac{0{,}35}{T_f}. \tag{7.2a, b}$$

For example, the rise time $T_n = 1$ ns and the descent time $T_f = 10$ µs correspond to the upper and lower limit frequencies of partial discharge pulses to $f_{mh} = 350$ MHz and $f_{md} = 35$ kHz.

Due to the existence of interference, the transmission of signals with this frequency band is practically impossible in the environment of electrical equipment, and therefore EN 60 207 recommends measuring the properties of partial discharges in three frequency bands: 30–100 kHz, 30–500 kHz,

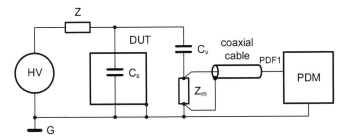

Figure 7.2   Block diagram of parallel connected object under test.

and 100–400 kHz. However, there are also *narrowband measurements* in the range of 9–30 kHz and *broadband measurements* in the range of 100–500 kHz.

7.1.2 Galvanic methods for measuring partial discharges

Galvanic methods are among the most widely used direct methods of measuring partial discharges, as they allow the most accurate determination of the sizes of individual charges of partial discharges and their derived parameters, including the time courses of the corresponding current pulses.

According to EN 60270, there are three basic connections for direct measurement of partial discharges. In all these connections, the tested insulation of the HV electric machine is excited by an AC voltage source with Z *impedance protection*.

The connection with a *parallel connected grounded object device under test* (DUT) has a coupling HV capacitor C_v connected with the coupling impedance Z_m parallel to the insulation system of the tested HV machine with a capacity C_s. Coupling impedance is connected by a coaxial cable with a partial discharge meter *partial discharge meter* (PDM) (Figure 7.2).

This most widely used method of wiring requires the connection of one terminal of the supply to the measured object with the neutral conductor of the three-phase power supply and its grounded skeleton. However, the disadvantage of wiring is the low sensitivity of partial discharge measurements.

For example, with the capacity of the insulation system of the object $C_s = 100$ nF, the capacity of the coupling capacitor $C_v = 1$ nF and the current pulses passing through the coupling capacitor C_v are reduced in a ratio of $1:10^2$ to the pulses passing through the insulation system of the measured object with the capacity C_s.

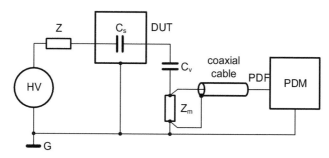

Figure 7.3 Block diagram of seriel connected object under test.

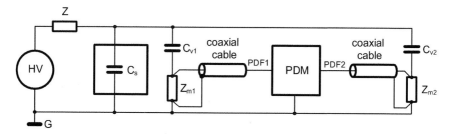

Figure 7.4 Differential connecting to measure partial discharges.

The connection with the *serially connected sensing impedance* with the insulation system of the test object according to Figure 7.3 is used to measure objects where it is not necessary to connect the isolated system with the ground.

Unlike the previous connection, this connection can achieve a higher sensitivity of partial discharge measurements, because the capacity of the insulation system of the test object is always significantly greater than the capacity of the coupling HV capacitor. If the dispersion capacities of the circuit to the ground are greater than the capacity of the measured object, the coupling HV capacitor C_v can be omitted.

The *differential connection* in Figure 7.4 with two coupling impedances Z_{m1} and Z_{m2} is mainly used in operational measurements, since this connection allows to effectively suppressing the influence of external interference. Signals from sensing impedances are measured by a two-channel PDM meter.

In addition, a capacitor with a capacity close to the capacity of the insulation system of the measured object C_s can be used as a coupling capacitor C_v. Coupling impedances Z_{m1} and Z_{m2} can also be frequency tunable and can

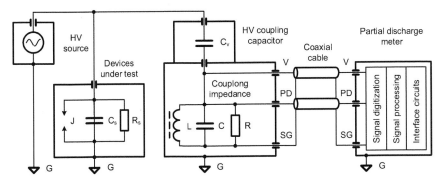

Figure 7.5 Detailed connecting to measure partial discharges of grounded HV machine.

be used to equalize the bridge with the aim of limiting or completely suppressing the influence of external interference on the accuracy of partial discharge measurements.

Detailed connecting of the system for measuring partial discharges with a grounded diagnosed HV electric machine is shown in Figure 7.5.

The AC HV source is usually made up of an HV autotransformer with a continuously variable dividing ratio. To suppress the distortion of the output voltage of the transformer, caused by the nonlinear dependence of the magnetization curve of their ferromagnetic cores, *passive LC filters* are connected to its output. With these sources, alternating voltages up to several tens of kilovolts can be generated at a current load of several units A with a distortion below 1%.

If the insulation system of the tested machine has a capacity C_s, then for the excitation current I_s and the required power of the source P_s

$$I_s = 2 \cdot \pi \cdot f \cdot C_c \cdot U_s, \qquad P_s = U_s \cdot I_s = 2 \cdot \pi \cdot f \cdot C_c \cdot U_s^2. \qquad (7.3a, b)$$

where U_s is the effective value of the test voltage and f is the mains frequency (50 Hz).

For example, excitation of the HV insulation system of an electric generator with a test voltage of 12 kV, whose capacity is 500 nF, requires generating an excitation current of 1.9 A at a main frequency of 50 Hz, which corresponds to an apparent power of 22.5 kVA.

Therefore, instead of AC test voltage, DC voltage has recently been used to measure partial discharges. The character and methods of evaluation of partial discharges in this method of power supply are similar to that of surface discharges of insulators.

Figure 7.6   OMICRON HV flush capacitors up to 50 kV.

7.1.2.1 Coupling capacitors

Coupling HV capacitors are produced in the range of hundreds of pF to nF units with nominal voltages in the range of 6 kV to 10 MV.

Coupling capacitors with a voltage of approximately 50 kV usually use an HV ceramic dielectric with a very high insulation resistance, greater than 1 TΩ, an extremely low *loss factor*, less than 10^{-6} A, and negligible partial discharges, less than 1 pC.

Figure 7.6 shows OMICRON HV coupling capacitors with a nominal voltage of up to 50 kV.

Coupling capacitors with higher nominal voltages have a dielectric consisting of sulfur hexafluoride SF_6 or compressed nitrogen and are usually equipped with a single or multiple *circular round electrodes* in the form of an anuloid to prevent *corona discharges*. For precise measurements of partial discharges, the coupling capacitors are shielded by external electromagnetic shielding connected to the ground of the measuring system.

Figure 7.7 shows HV coupling capacitors 100 pF and 1 nF from PHOENIX for voltages up to 500 kV.

7.1.2.2 Coupling impedance

They are designed to filter out the excitation alternating voltage, reaching tens of kilovolts, to the level of several volts, to limit the frequency range of measured partial discharges in the recommended frequency range, to impedance adjustment of the connection of partial discharge meters and to protect them from extreme HV interfering pulses that may occur when measuring partial discharges.

Figure 7.8 gives an example of the connection of a *simple coupling impedance* with the corresponding frequency characteristic of its transmission for the PDF output to the partial discharge meter connection.

a) with a nominal voltage to 100 kV   b) with a nominal voltage to 500 kV

Figure 7.7   Coupling HV capacitors from PHOENIX.

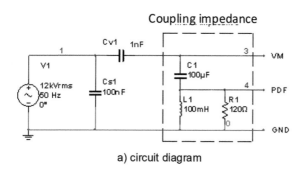

a) circuit diagram

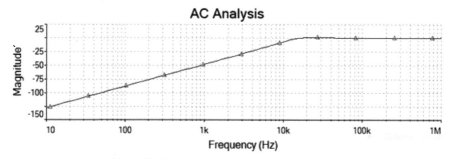

b) amplitude frequency response of PDF output

Figure 7.8   Coupling impedance with single LC filter.

The coupling impedance consists of a *second-order high pass* with a choke L_1 and capacitor C_2 and a damping resistor R_1. The wiring diagram is complemented by an excitation AC source V_1, a capacitor C_s, representing the capacity of the insulation system of the diagnosed object, and an HV

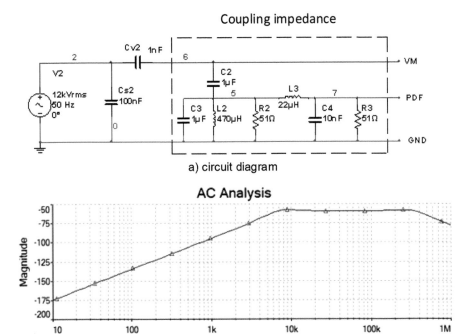

a) circuit diagram

b) amplitude frequency characteristics of PDF output binding impedance

Figure 7.9   Coupling impedance with fourth order filter.

coupling capacitor C_v. PDF output is intended for connection to the partial discharge meter.

From the course of the amplitude frequency characteristic of the transmission of this connection, it is clear that the AC excitation voltage of 12 kV with a frequency of 50 Hz by 100 dB, i.e., at 12 mV, is evident. This simple coupling impedance is therefore not suitable for measuring partial discharges with low amplitude and, moreover, does not limit their upper limit frequency.

Improved engagement of the coupling impedance with a fourth-order high-pass filter, and with the corresponding amplitude frequency response of its transmission, is shown in Figure 7.9.

The coupling impedance consists of a second-order high pass with an L_2 choke and a C_2 capacitor with a damping resistance of R_2. A second-order low-pass filter consisting of the L_3, choke, and capacitor C_4 with damping resistor R_3, whose resistance of 50 Ω is equal to the wave impedance of the coaxial cable through which the *coupling impedance* is connected to the partial discharge meter, is connected to this high-pass filter.

Figure 7.10 OMICRON CPL543 coupling impedance.

The transmission of pulses without their reflections is conditioned by the same input impedance (50 Ω) of the partial discharge meter. The VM output is designed to measure the HV excitation voltage, and the PDF output is designed to measure partial discharges.

From the courses of the amplitude frequency characteristic of the coupling impedance, it is clear that the excitation HV voltage of 12 kV with a frequency of 50 Hz is reduced by more than 140 dB, i.e., to 1.2 mV, which makes it possible to measure even small partial discharges. In addition, this connection suppresses signals with frequencies above 300 kHz.

The coupling impedance is produced either in the form of a separate device or they can be part of the HV coupling capacitor at lower test voltages.

Figure 7.10 shows the coupling impedance CPL 543 from OMICRON, with a filter capacitor capacity of 10 μF and a maximum passing current of 4 A. Sockets with a diameter of 4 mm are used to connect the HV coupling capacitor. PD and HV outputs are designed to measure partial discharges and to measure HV test voltage with a partial discharge meter.

Figure 7.11 shows the design of the coupling capacitor 1 nF CCI-30DC with built-in coupling impedance, made at the Department of Measurement at the Faculty of Electrical Engineering of the Czech Technical University in Prague.

Since the capacity of the HV coupling capacitor with a ceramic dielectric decreases significantly with increasing stress, this dependence is corrected when determining the effective value of the excitation voltage (Figure 7.12).

This design solution, usable in test voltages of tens of kilovolts, eliminates the need for external connection of the coupling capacitor with the coupling impedance input, which more effectively eliminates the influence of interfering signals in the transmission of pulses that arise during the measurement of partial discharges and, in addition, eliminates the risk of the existence of HV voltage in the event of a disconnected connection of the HV capacitor with the coupling impedance.

Figure 7.11   Coupling capacitor with built-in
coupling impedance.

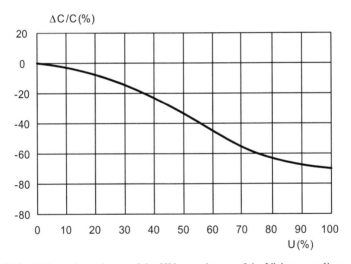

Figure 7.12   Voltage dependence of the HV capacitance of the Vishay coupling capacitor.

7.1.3 Measurement of partial discharges

Oscilloscopes were originally used to measure partial discharges, which allowed to depict the time course of the occurrence of partial discharges, including the time course of alternating test voltage. *Modern devices* digitize these waveforms with fast ADCs using signal processors or gate arrays to process the results.

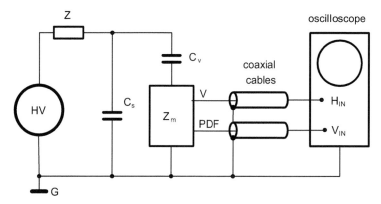

Figure 7.13 Measurement of partial discharges with an oscilloscope.

7.1.3.1 Measurement of partial discharges with oscilloscope

If the time base of the oscilloscope is synchronized by the frequency of the test voltage, then pulses that are modulated on the sinusoidal course of the test alternating voltage represent the time courses of partial discharges.

If the horizontal inputs of the H_{IN} oscilloscope are connected to the voltage outputs V of the measuring impedance and its vertical V_{IN} inputs are connected to the output of the PDF of the measuring impedance, then the time courses of the partial discharges are displayed in the so-called *elliptical view* (Figure 7.13).

The connection between the two types of partial oscilloscope discharge imaging is shown in Figure 7.14.

The advantage of this method is in its principled simplicity; the disadvantage is that it is not applicable to accurate qualitative evaluation of partial discharge parameters, including further processing of measured results.

In Figure 7.15, examples of partial discharge display in sinusoidal and elliptical modes of display are shown.

7.1.3.2 Measurement of partial discharges by a peak detector

A partial discharge meter with a *peak detector* usually has two separate channels; one channel is designed to measure partial discharges, and the other channel is designed to measure the RMS value of the alternating excitation voltage (Figure 7.16).

The channel for measuring partial discharges consists of a surge protector, an amplifier with automatically switchable amplification, allowing optimal use of the distinguishability of the A/W converter in terms of digitization of voltage pulses from the coupling impedance, a top detector with an adjustable detection level, and an A/D converter.

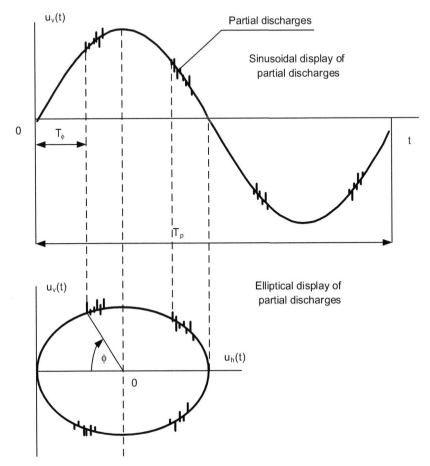

Figure 7.14   Relationship between sinusoidal and elliptical imaging of partial discharges by oscilloscope.

The *peak detector* uses a *memory capacitor C* to capture peak pulse signal values, which is charged via a fast *Schottky diode* (SD) by the output current of the OA_1 operational amplifier. The OA_2 operational amplifier is connected in the function of a voltage monitor with +1 transmission with high input impedance, preventing the discharge of the memory capacitor charge when the pulse signal voltage peak is captured (Figure 7.17).

The time waveforms of the signals in the partial discharge meter with a peak detector are given in Figure 7.18.

If the voltage pulses at the input of the amplifier are greater than the set threshold voltage U_{p0}, then the voltage pulse amplitudes are switched on for

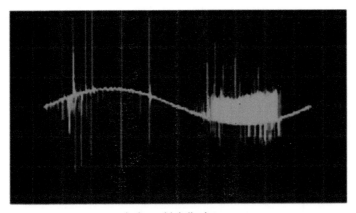

a) sinusoidal display

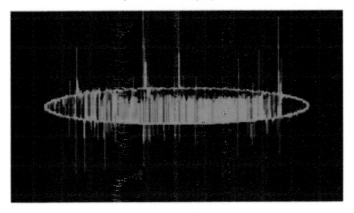

b) elliptical display

Figure 7.15 Ways of displaying partial discharges with an oscilloscope.

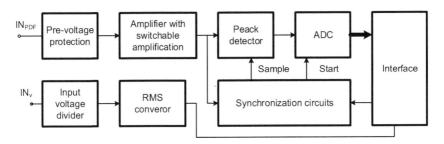

Figure 7.16 Partial discharge meter with peak detector.

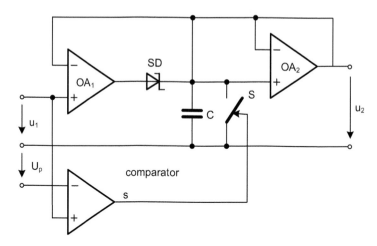

Figure 7.17   Connection of a peack detector with zero switch.

the duration of the *acquisition time* T_a sample at the H level of the $s(t)$ signal, and then the amplitudes of the voltage pulses are converted to equivalent digits for the duration of the *conversion time* T_c. Subsequently, the switch is switched on and the charges on the memory capacitor are discharged. If the amplitudes of the pulse signals do not exceed the set threshold voltage U_p, then they will not be captured and subsequently converted.

If more pulse signals with different amplitudes appear during the sampling time, then the largest amplitude of these signals is always digitized. Obviously, with this method of measuring peak values, impulse signals cannot be indicated during the conversion of the A/D converter, and, therefore, this time is referred to as *dead time*.

When using fast operational amplifiers with transit frequencies of hundreds of megahertz and switching *Schottky diodes* with switching times of several nanoseconds, sampling times of several tens of nanoseconds can be achieved, and when using A/D converters with a resolution of 12 bits with sample rates of several MSps (*samples per seconds*), dead times can be reduced to μs units. The accuracy of the determination of amplitudes reaches several percentages for these meters.

To measure the *root mean square* (RMS) value of the excitation voltage, these meters use logarithmic converters of the RMS, whose output DC voltage is digitized by A/D converters with a resolution of 12–14 bits.

7.1.3.3 Partial discharge meters with direct signal digitization

To digitize these signals, they use fast A/D converters with a resolution of 12–16 bits with sampling rates of several GSps, which corresponds to the

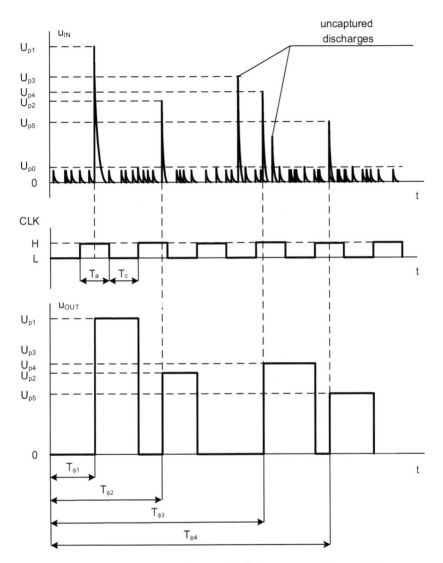

Figure 7.18 Signal time courses in a partial discharge meter with a peak detector.

time resolution of sampling several nanoseconds. With these partial discharge meters, the frequency spectrum of the reconstructed signal is theoretically limited by half the sampling rate, which corresponds to the frequency range of several hundred megahertz.

This limitation is not a problem, because the recommended frequency ranges when measuring partial discharges are at most tens of megahertz, but when calculating the charge from such digitized pulses, errors arise due to the staircase approximation of digitized pulses (Figure 7.19).

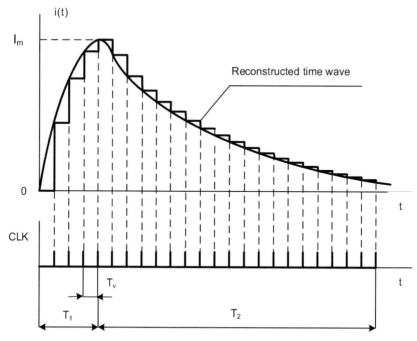

Figure 7.19   Time course of time-discretized pulse signal.

The time course of the measured current pulse about the amplitude I_m with the *rise time* T_r and the *fall time* T_f by the following equation:

$$i(t) = I_m \left[1 - e^{-\frac{t}{\tau_n}} + e^{-\frac{t-T_1}{\tau_f}} \right], \tag{7.4}$$

where $\tau_n = 0.35 T_n$ is the *rise time constant* and $\tau_f = 0.35 \cdot T_f$ is its *fall time constant*, and the time-discretized current pulses defined by the following equation:

$$i_{di}(t) = I_m \left[1 - e^{-\frac{t_i}{\tau_n}} + e^{-\frac{t_i-T_1}{\tau_f}} \right], i = 1, 2, ..., \tag{7.5}$$

and then the difference in the surfaces below these waveforms determines the time discretization error, which increases with a decreasing T_1/T_v ratio, where T_1 is the time that is approximately equal to the *rise time* T_n and T_v is the period of the sampling signal.

For example, if a current pulse with an in-rush time of 10 ns sampling rate of 1 GSps is digitized, the time discretization error is approximately 5%. If the rise time is 100 ns, then at the same sample rate of 1 GSps, the time

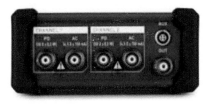

a) front panel   b) back panel

Figure 7.20   MPD600 partial discharge meter by OMICRON.

discretization error drops below 1%. The digitized signal is recorded into the operating memory of the device and further processed in a gate array, including digital filtration of the signal.

To measure the RMS value of the test voltage, these meters also use direct digitization of the excitation signal, the effective value of which is determined by calculation according to the definition relationship

$$U_{\mathrm{RMS}} = \sqrt{\frac{1}{n}\sum_{i=1}^{n}u_i^2}, \tag{7.6}$$

where n is the number of voltage samples taken from the measured course.

For example, using a 12-bit A/D converter with a sample rate of 200 kHz, an accuracy of more than 0.1% can be achieved to determine the effective value of the sinusoidal waveform of the test voltage.

The advantage of partial discharge meters with direct digitization of pulse signals is not only to achieve significantly higher accuracy of determination of partial discharge charges but also the possibility of further processing and presentation of measured data, including the determination of the frequency of their occurrences.

A typical partial discharge operating meter using direct digitization of pulse signals is, for example, the MPD600 meter from OMICRON, allowing to measure partial discharges in the frequency range of 0–20 MHz with switchable limit frequencies of filters of 9, 40, 100, 160, 300, and 650 kHz and 1 and 3 MHz in accordance with EN 60 270. The channel for measuring partial discharges has a time resolution of less than 1 ns.

The dynamic range of the meter is greater than 100 dB, i.e., $1{:}10^5$ and a resolution of 1 pC. The achievable accuracy of partial discharge measurement is 1% and the effective value of the excitation voltage is 0.05%.

A more modern two-channel variant of this meter is the MPD650 meter, allowing simultaneous measurement of discharges in two channels. Both meters have isolated data transmission to the control unit via optical cable and also have a battery power supply (Figure 7.20).

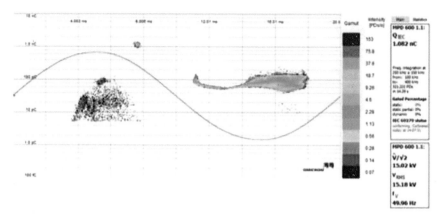

Figure 7.21 Time course of partial discharges in air.

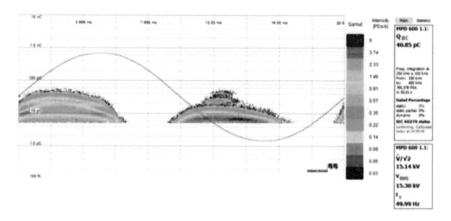

Figure 7.22 Time course of partial discharges in oil.

Figures 7.21–7.23 show examples of measurements of partial discharge waveforms in air, oil, and epoxy dielectrics on ball electrodes, with higher frequencies of their occurrence being expressed in warmer colors.

A similar device PDA100 was developed at the Department of Measurement of the Faculty of Electrical Engineering of the Czech Technical University in Prague (Figure 7.24).

This partial discharge meter is also *two-channel* and allows partial discharges in the range of 10 pC to 100 nC in the frequency range of 30–300 kHz, while the setting of the limit frequencies of the filters is also in accordance with EN 60 270.

The meter contains two identical channels with 16-bit A/D converters and sample rates of up to 500 MSps. A gate array is used for digital

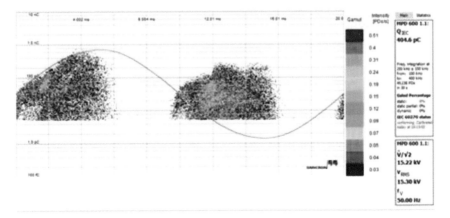

Figure 7.23 Time course of partial discharges in a solid dielectric.

a) front panel   b) back panel

Figure 7.24 Partial discharge meter PDA100 by CTU in Prague.

processing of the measured data, in which signal filtration, pre-processing, and transmission to a computer via an optical cable with a transmission rate of up to 1 Gbit/s are implemented. The device is also powered by batteries with an operating time of 10 hours.

Figure 7.25 shows the user interface of the PDA100 partial discharge meter.

Figure 7.26 shows the frequency of occurrence of partial discharges when measured by the PDA100 partial discharge meter on a 12 kV/980 kVA HV engine.

Elimination of interfering signals can be achieved in most digital partial discharge meters using vertical and horizontal signal gate.

Horizontal gating can eliminate interfering signals that occur at the same place of the measured waveform. Gatekeeping can usually be done either in one period or in two half-periods of the course.

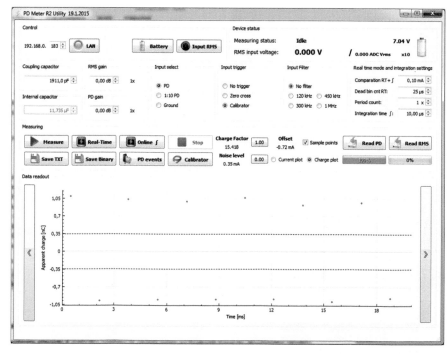

Figure 7.25   Partial discharge meter user interface PDA100.

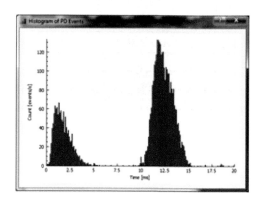

Figure 7.26   Frequency of occurrence of partial discharges.

Figure 7.27 shows the waveforms of the test AC voltages in an elliptical display using horizontal gate.

Figure 7.28 shows the time course of the test voltage when using horizontal and vertical gates. Measured pulses are marked in green, time-gated pulses in red, pulses exceeding the specified level are displayed in blue, and amplitude voltage levels that are not evaluated are displayed in purple.

a) single gate   b) double gate

Figure 7.27 Elliptical display of alternating test voltages using horizontal gates.

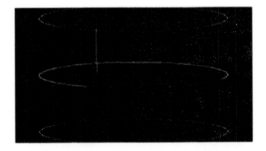

Figure 7.28 Elliptical display of alternating test voltages under gate.

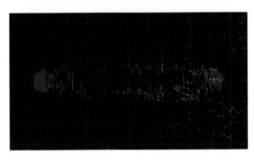

Figure 7.29 Elliptical display of noise voltage with set levels of vertical gate.

An example of an elliptical display of noise voltage, the amplitude of which is the range of set levels of vertical gate, is shown in Figure 7.29.

7.1.4 Charge calibration

Because when measuring partial discharges, the capacity of the insulation system of the measured object is never known, affecting the transfer of partial discharge charges to the partial discharge meter, it is always necessary to perform a charge calibration of the entire measuring chain before each measurement. For this purpose, *charge calibrators* are used, which generate defined charges to the measuring circuit when the HV source is disconnected.

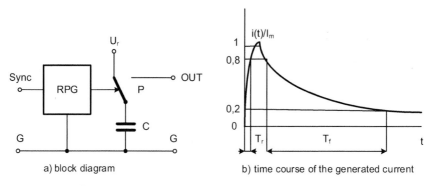

a) block diagram
b) time course of the generated current

Figure 7.30   Principe of connection of the charge calibrator.

Figure 7.31   Cfarge calibrator CAL 542
by OMICRON.

The principle of connection of the charge calibrator is shown in Figure 7.30.

A *charge calibrator* consists of a synchronized *rectangular pulse generator* (RPG), the output signal of which is controlled by switch S and switching capacitor C to either the reference voltage source U_r or to the output of the calibrator. The product of the reference voltage with the capacitance of the capacitor then determines the charge of the current signal generated in this way into the HV insulation system of the electric machine

$$Q = C \cdot U_r. \tag{7.7}$$

Decimal ranges of the generated charge are usually solved by switching capacitors with decimal graded values such as 100 pF and 1, 10, or 100 nF.

According to the recommendations of EN 60 270, the *rise time* of the current pulses of the calibrator $T_r < 100$ ns and the *fall time* $T_f < 10$ ms at its nominal load by a *non-inductive resistor* of 100 Ω.

Figure 7.31 shows OMICRON's CAL 542 cartridge calibrator, allowing to generate current pulses with charges graduated in four decimal ranges of 10 and 100 pF and 1 and 10 nF with rise times of less than 4 ns and a maximum repetition rate of 300 Hz.

Figure 7.32 Charge calibrator PDC2 by CTU in Prague.

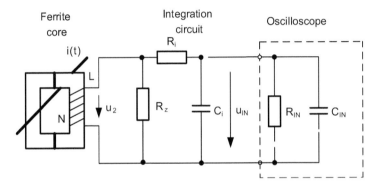

Figure 7.33 Connecting the current transformer to the oscilloscope.

Figure 7.32 shows the PDC2 cartridge calibrator, developed at the Department of Measurement of CTU FEE in Prague. This calibrator allows you to generate 2–16 current pulses in a period of 20 ms in the charge range of 10 pC to 100 nF with an accuracy of 1%. The *rise time* of the generated pulses is less than 50 ns.

7.1.5 Measurement of partial discharges by current transformers

The current transformer consists of a *toroidal closed core* of magnetically soft material (*ferrite*) through which the power wire of the diagnosed HV electric machine is threaded (Figure 7.33).

A telecoil of N turns is wound on the core of the transformer, in which the voltage is induced, given by the change in the magnetic flux in the core $d\Phi/dt$

$$u_2\left(t\right) = N \cdot \frac{d\Phi}{dt} = \frac{N}{R_m} \cdot \frac{di}{dt}, \qquad (7.8)$$

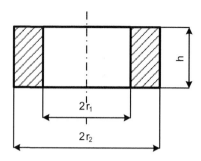

Figure 7.34   Toroidal ferrite transformer core with rectangular cross section.

where di/dt is the time changes in the current in the HV supply wire of the electric machine, and

$$R_m = \frac{l_m}{\mu S_m},$$
(7.9)

is the magnetic resistance of the core, given by its permeability μ, the mean length of the field line in the core l_m, and its cross-section S_m. Inductance of the telecoil of N turns is

$$L = \frac{N^2}{R_m},$$
(7.10)

where R_m is the magnetic resistance of the transformer.

If the ferrite core of the transformer has a rectangular cross-section of inner radius r_1, outer radius b, and height h, then for the inductance of the coil (Figure 7.34),

$$L = \frac{\mu \cdot h \cdot (r_2 - r_1) \cdot N^2}{2\pi r_1}.$$
(7.11)

This inductance forms with the load resistance R_z, a derivative cell with a time constant $\tau_h = L/R_z$, which determines the *higher limit frequency* of transmission of the current transformer

$$f_h = \frac{1}{2 \cdot \pi \cdot \tau_h} = \frac{R_z}{2 \cdot \pi \cdot L}.$$
(7.12)

If an integration cell with the time constant $\tau_i = R_i C_i$ is connected to the load resistance, the *lower limit frequency* of transmission of the current transformer is

$$f_d = \frac{1}{2 \cdot \pi \cdot \tau_d} = \frac{1}{2 \cdot \pi \cdot R_i \cdot C_i}.$$
(7.13)

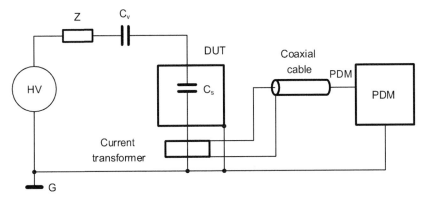

Figure 7.35 Location of the current transformer when measuring partial discharges.

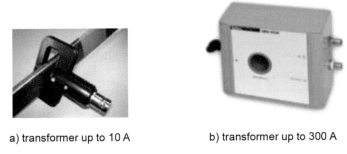

a) transformer up to 10 A   b) transformer up to 300 A

Figure 7.36 Current transformers for partial discharge measurement.

For example, at coil inductance $L = 200$ nH and load resistance $R_z = 100\ \Omega$, the higher limit frequency of the transformer is $f_h = 3$ MHz, and at $R_i = 1\ M\Omega$ and $C_i = 60$ pF, the *lower limit frequency* of the probe is 9 kHz. This frequency range corresponds to the frequency range of most produced partial discharge meters.

Thread transformers are usually placed on supply wires near their zero potential (Figure 7.35).

Figure 7.36 shows examples of designs of current beam transformers designed for non-contact measurement of partial discharges.

7.1.6 Measurement of partial discharges by inductive probes

In terms of design and use, we distinguish between single, grooved, and differential inductive probes.

A *simple inductive probe* consists of an air coil into which, due to changes in the magnetic field, provoked by the current pulses of partial

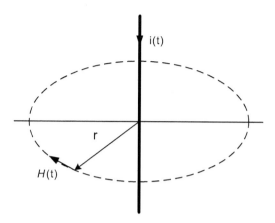

Figure 7.37   Simple inductive probe in the field of the electromagnetic field of the conductor.

discharges, voltage pulses are induced, which, after integration, determine the charge of partial discharges.

If a conductor through which a time-varying current $i(t)$ flows is located at a distance r from the coil, then this current causes a time change in the intensity of the magnetic field $dH = di(t)/r$ at a distance r from the conductor, and in a coil of N threads and area S, perpendicular to the voltage intensity phasor, a voltage is generated

$$u(t) = N\frac{d\Phi(t)}{dt} = \frac{N \cdot S \cdot \mu_0}{r}\frac{di(t)}{dt}, \qquad (7.14)$$

where $d\Phi/dt = \mu.S.dH/dt$ is the time change in the magnetic flux encroaching on the coil, μ is the permeability of air, which is approximately equal to the permeability of vacuum $\mu_0 = 4\pi.10^{-7}$ [H/m] (Figure 7.37).

If the integration cell $R_1 C_1$ is connected to the coil, then at its output, there is a voltage

$$u_{1i}(t) = \frac{1}{R_1 \cdot C_1}\int u(t) = \frac{N \cdot S \cdot \mu_0}{r}i(t). \qquad (7.15)$$

If this voltage is once again integrated with the time constant $R_2 C_2$, then the output voltage of the second integrator is directly proportional to the charge of the pulses that pass through the conductor

$$u_{2i}(t) = \frac{1}{R_2 \cdot C_2}\int u_{1i}(t)dt = \frac{1}{R_1 \cdot C_1 \cdot R_2 \cdot C_2}\frac{N \cdot S \cdot \mu_0}{r}q. \qquad (7.16)$$

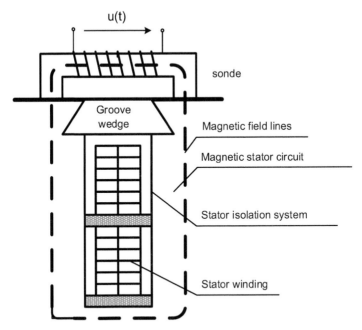

Figure 7.38 Method of applying the inductive probe to the winding groove.

A prerequisite for the plausibility of the determination of partial discharge charges by this method is that the time constants of both integrators are always several orders of magnitude larger than the time constant corresponding to the lower limit frequency f_d of the partial discharge measurements, i.e., that $R_1 C_1 \gg 1/(2 \cdot \pi \cdot f_d)$ and $R_2 \cdot C_2 \gg 1/2 \cdot \pi \cdot f_d$ apply. The high limit frequency of the probe is limited by the resonant frequency $f_o = 1/2 \cdot \pi \cdot (L_c \cdot C_c)^{-1/2}$, where L_c is the inductance, L_c is the inductance, and C_c is the coil's own capacity.

The disadvantage of this probe is its low sensitivity, decreasing with increasing distances of the probe from the place of formation of partial discharges and its dependence on the permeability of the environment in which the place of formation of partial discharges is located. Another disadvantage of this solution is low resistance to disturbing alternating magnetic field.

To diagnose the stator windings of HV rotating electric machines, in which the rotor of the machine is removed, contact probes are used, which are formed by a *ferrite core* with a wound coil. These probes are applied above the groove of the stator winding of the machine and, together with its magnetic circuit, form a current transformer in the winding of which *current pulses* are induced, caused by partial discharges in the machine winding (Figure 7.38).

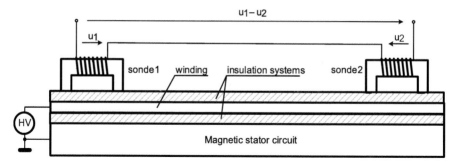

Figure 7.39   Measurement of partial discharges with differential electromagnetic probe.

Differential electromagnetic probes are mainly used to locate the occurrence of partial discharges in the stator grooves of HV electric rotating machines. The probe consists of two inductive probes, each located at one end of the stator groove and whose probe windings are electrically connected opposite to each other (Figure 7.39).

Under the influence of the electromagnetic field, which is provoked by currents of partial discharges, pulses from the point of the source of discharges propagate in both directions. If the current pulse of the partial discharge passes through both probes in the same direction, the output voltages of the probes have the *opposite polarity* and are therefore subtracted, which means that the place of occurrence of partial discharges lies outside the part that is bounded by the probes.

If the location is between the probes, the probe signals have the same polarities and their induced voltages are summed up. In addition, this connection eliminates the influence of external interference and the influence of discharge activity in adjacent grooves.

7.1.7 Measurement of partial discharges by capacitive probes

Simple, grooved, and coaxial capacitive probes are used to measure and localize the occurrence of partial discharges.

A *simple capacitive probe* consists of a circular electrode that is attached in isolation to the probe housing, in the front of which there is a cover of non-conductive material, most often made of glass (Figure 7.40).

By the action of an *AC electrostatic field* near the sensing capacitor C_s, caused by partial discharges, the charge q is induced on the electrode of the sensing capacitor, which is converted by the *charge amplifier* into voltage

$$u(t) = \frac{q}{C_i},$$
(7.17)

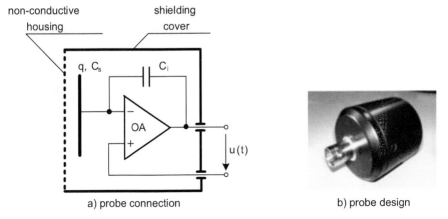

a) probe connection   b) probe design

Figure 7.40 Single capacitive probe.

where C_i is the integration capacity of the charge amplifier.

The disadvantage of the probe is its low sensitivity, decreasing with increasing distances of the probe from the place of formation of partial discharges and its dependence on the permittivity of the environment, which is located between the place of formation of partial discharges and the probe. Another disadvantage is its low resistance to a disturbing alternating electrostatic field.

The *capacitive groove probe* consists of an insulating laminate board, on which a shielding copper layer is placed on one side, and on the other side of the board, there is a sensing electrode in the form of a narrow copper strip. The probe is inserted into the stator grooves of the machine under the groove wedges.

Figure 7.41 shows the design of the slotted probe and its location in the stator groove of the rotary electric machine.

For the capacity between the shielding and sensing electrode of the probe,

$$C = \varepsilon_0 \cdot \varepsilon_r \frac{wl}{d}, \tag{7.18}$$

where $\varepsilon_0 = 8.854 \cdot 10^{-12}$ [F·m^{-1}] is the vacuum permittivity, $\varepsilon_r \approx 2$ is the relative permittivity of the laminate, d is its thickness, and w and l are the width and length of the sensing electrode, respectively.

If a partial discharge q occurs near the capacitive probe, then a voltage is created between the probe electrodes, which is proportional to the charge q of this discharge

$$u(t) = k\frac{q}{C}, \tag{7.19}$$

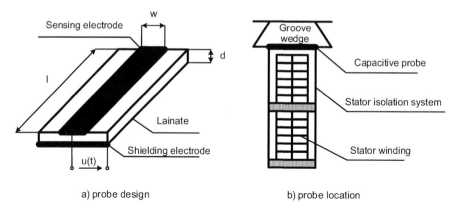

Figure 7.41   Grove capacitive probe.

where $k < 1$ is the coefficient of charge transfer from the point of its origin to the capacitor dielectric.

The *coaxial capacitive probe*, suitable for detecting partial discharges in HV cables, consists of a coaxial capacitor, for whose capacities between the inner conductor and the inner jacket C_1 and the capacities between the inner and outer shielding jacket C_2 apply

$$C_1 = \frac{2 \cdot \pi \cdot \epsilon_1}{\ln \frac{d_2}{d_1}}, \qquad C_2 = \frac{2 \cdot \pi \cdot \epsilon_2}{\ln \frac{d_3}{d_2}}, \qquad (7.20a, b)$$

where d_1 is the diameter of the inner conductor, d_2 and d_3 are the inner diameters of the inner and outer shell of the probe, l is the probe length, and ϵ_1, ϵ_2 are the permittivities of the environment between the inner conductor and the inner jacket, and between the inner and outer shielding jackets (Figure 7.42).

If a partial discharge q occurs in a dielectric between the inner conductor and the inner jacket, then this charge causes a voltage between the two shells, which is proportional to the charge of this discharge

$$u(t) = \frac{C_1}{C_1 + C_2} q. \qquad (7.21)$$

Capacitive probes enable relatively precise localization of the source of partial discharges, especially in HV cables. Their disadvantage is the impossibility of determining the source of partial discharges of insulation systems that are electrically shielded, because then it is impossible to measure the electric field emitted into the surroundings.

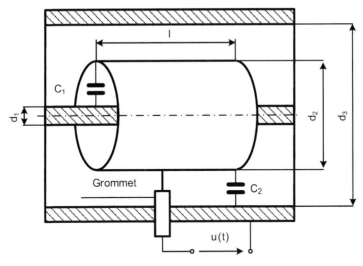

Figure 7.42 Coaxial capacitive probe.

7.2 Non-electrical Methods for the Detection of Partial Discharges

Non-electrical manifestations of partial discharges include sound effects in the *audible* and *ultrasonic bands*, *light radiation* in the visible and *ultraviolet spectrum*, and the formation of *chemical fumes*. With the help of these manifestations, using appropriate methods, it is possible not only to detect but also to quantify partial discharges in the insulation systems of HV electrical machines.

7.2.1 Ultrasonic detection of partial discharges

Ultrasonic piezoelectric sensors are used for the acoustic detection of partial discharges, operating most often in the frequency band of 20–200 kHz, since it is in this band that partial discharges manifest themselves most strongly.

Piezoelectric sensors use the so-called *direct piezoelectric effect*, in which the mechanical deformation of the crystals of some dielectrics leads to polarization of their dipoles, which causes the formation of a charge on the crystal electrodes, the size of which is proportional to the mechanical stress of the crystal. If the force F acts in the direction of the electric y-axis of the crystal cut, it is a transverse piezoelectric effect in which the polarization vector is parallel to the x-axis. For the charge generated on the crystal electrodes,

$$q = k_d \frac{b}{a} F_y, \tag{7.22}$$

where k_d is the *piezoelectric constant of the dielectric* and a and b are its dimensions of its cut.

If this crystal cut is connected to an *integration amplifier* with an integration capacitor C_i, then its output voltage is directly proportional to the generated charge and therefore to the pressure $p = F/(a.b)$, acting on this cut

$$u = -k_d \cdot \frac{b^2}{C_i} \cdot p. \tag{7.23}$$

The capacity of the sensor is determined by its dimensions and by permittivity e of the crystal

$$C_s = \varepsilon \cdot \frac{a \cdot c}{b}. \tag{7.24}$$

This capacity and the capacity of the supply cable practically do not affect the transmission of the integrator, since both capacities are connected to the inverting input of the operational amplifier. The voltage at this input is practically negligible with respect to the signal country, assuming a large differential amplification of the operational amplifier.

Figure 7.43 shows the cut of the crystal and the method of its connection to the integration amplifier.

The most commonly used piezodielectrics are silicon oxide SiO_2, barium titanate $BaTIO_3$, and lead zirconicanite $PbZrO_3$. For example, the SiO_2 crystal has a piezoelectric constant $k_d = 2.3 \cdot 10^{-12}$ [C/N^{-1}].

Ultrasound sensors are used in either *single* or *differential arrangements* (Figure 7.44).

In *differential arrangement*, for distances l_1 and l_2 of sensors S_1 and S_2 from the place of origin of partial discharges

$$l_1 = \sqrt{d^2 + s_1^2}, \qquad l_2 = \sqrt{d^2 + s_2^2}. \tag{7.25a, b}$$

Assuming that ultrasonic waves propagate to the sensors at different propagation rates v_1 and v_2, then for the ratio of their impact times on the sensors,

$$\frac{T_1}{T_2} = \frac{l_1}{v_1} \frac{v_2}{l_2}. \tag{7.26}$$

If these times are identical, then the axis of the sensors is identical to the axis of the place of origin of partial discharges. This knowledge can be used to find the place of origin of partial discharges.

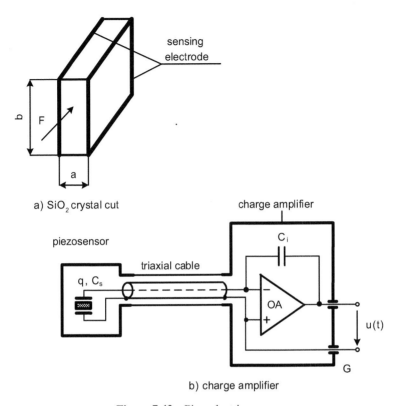

a) SiO$_2$ crystal cut

b) charge amplifier

Figure 7.43   Piezoelectric sensor.

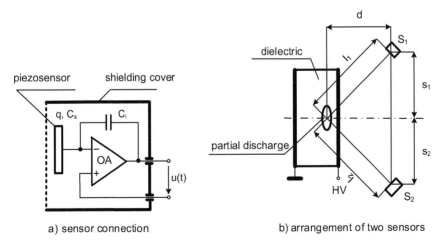

a) sensor connection

b) arrangement of two sensors

Figure 7.44   Ultrasonic partial discharge sensor.

In gases, the propagation rate of ultrasonic waves is determined by their pressure p, density ρ, temperature θ, and coefficient of thermal expansion v

$$v\left(\text{gass}\right) = \left(1 + \frac{\vartheta \, \upsilon}{2}\right) \cdot \sqrt{k \frac{p}{\rho}}. \tag{7.27}$$

For the airborne environment, the propagation rate is defined by the following equation:

$$v\left(\text{air}\right) = 331.82 + 0.606 \cdot \vartheta \left\lceil \text{ms}^{-1} \right\rceil. \tag{7.28}$$

In liquids, the propagation rates of ultrasonic waves are determined by the *modulus of volume elasticity K* and the density of the liquid ρ. In solid materials, then the *Young modulus of elasticity in tension E* and the *density* of the material ρ

$$v\left(\text{liquid}\right) = \sqrt{\frac{K}{\rho}}, \qquad v\left(\text{solid}\right) = \sqrt{\frac{E}{\rho}}. \tag{7.29}$$

For example, at 30°C, the propagation rate of ultrasonic waves in air and nitrogen is approximately 347 ms^{-1}, in water 1440 ms^{-1}, in transformer oils 1560 ms^{-1}, in glass-textite insulation 2630 ms^{-1}, in tempered paper 2220 ms^{-1}, in copper and aluminum 5100 ms^{-1}, and in steel 3520 ms^{-1}.

Figure 7.45 shows signal delays from two partial discharge ultrasonic sensors.

The advantage of using ultrasonic sensors is their easy applicability without the need to intervene in diagnosed HV machines. Their disadvantage is considerably sensitive to ambient noise, which can be partially suppressed by appropriate filtration.

A typical handheld partial discharge meter is, for example, the PD-SGS meter from Baur, containing both capacitive and ultrasonic sensors. The device allows you to measure partial discharges not only in real time but also their recording for 5 s (Figure 7.46a).

DOBBLE LEMKE's LDP-5 allows it to detect partial discharges with a resolution of less than 1 pC up to a distance of several meters in the frequency range up to 100 kHz (Figure 7.46b).

7.2.2 High-frequency detection of partial discharges

It determines the properties and locations of partial discharges based on the detection of the high-frequency *ultra-high frequency* (UHF) electromagnetic field generated by the discharge.

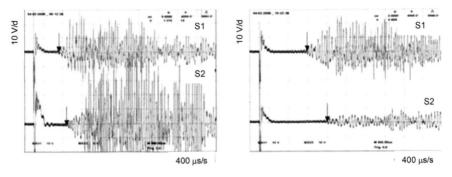

Figure 7.45 Delay of signals from two partial discharge ultrasonic sensors.

a) PD-SGS meter from Baur   b) LDP-5 meter from DOBBLE LEMKE

Figure 7.46 Partial discharge detectors.

In partial discharges, the electrons that carry the charge are torn off for a short moment from the atoms and molecules of the material and subsequently returned to their original place. This phenomenon generates a high-frequency electromagnetic field with an intensity of E, which propagates to the surroundings with a permittivity ε in all directions with an intensity inversely proportional to the square of the distance r from the place of origin of partial discharges (*Coulomb's law*)

$$E = \frac{Q}{4 \cdot \pi \cdot \varepsilon \cdot r^2},\tag{7.30}$$

where Q is the charge of the partial discharge.

The *propagation rate* of electromagnetic waves is defined by the permittivity ε and permeability μ of the environment through which these waves propagate

$$v = \frac{1}{\sqrt{\varepsilon \cdot \mu}}.\tag{7.31}$$

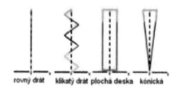

Figure 7.47   UHF probes for measuring partial discharges.

The rate of propagation of electromagnetic waves in air is approximately equal to the speed of propagation of these waves in a vacuum since its permittivity and permeability are approximately equal to the permittivity and permeability of the vacuum ε_0 and μ_0

$$v_o = \frac{1}{\sqrt{\varepsilon_0 \cdot \mu_0}} \approx 3 \cdot 10^8 \left[\frac{m}{s}\right]. \tag{7.32}$$

For example, at a temperature of 30°C, the rate of propagation of electromagnetic waves in water is $2.67 \cdot 10^8$ m/s, transformer oils $2.12 \cdot 10^8$ ms^{-1}, glass-textite insulation $1.34 \cdot 10^8$ ms^{-1}, and hardened paper $1.09 \cdot 10^8$ ms^{-1}.

The resulting shock waves are scanned by a UHF probe consisting of UHF antennas, whose signal is amplified by UHF amplifiers with GaS transistors. According to IEC 62 478, the frequency range of detection of the high-frequency field generated by partial discharges is 300 MHz to 3 GHz.

The most widely used UHF probes used to detect partial discharges are wire or disc probes in single or conical design (Figure 7.47).

For the application of these probes, technological holes are used in the covers of HV machines, e.g., transformers, or these probes are attached outside their covers, which are equipped with non-conductive windows (Figure 7.48).

When using three independent UHF probes S_1, S_2, and S_3, located, e.g., inside the transformer or on its non-conductive part of the surface, it is possible to determine the place of occurrence of partial discharges from the time differences of ΔT_i, $i = 1, 2, 3,...$ of their detected voltages

$$\Delta T_i = \frac{1}{v}\sqrt{\left(x_i - x_0\right)^2 + \left(y_i - y_0\right)^2 + \left(z_i - z_0\right)^2}, = 1, 2, 3,... \tag{7.33}$$

Figure 7.48 Location of UHF probes on the diagnosed transformer.

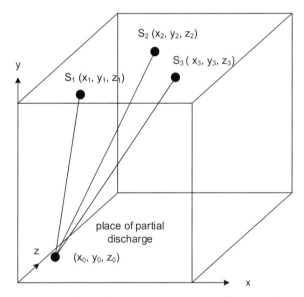

Figure 7.49 Determination of the place of origin of partial discharges by means of UHF probes.

where v is the speed of propagation of electromagnetic waves in a given environment (Figure 7.49).

Figure 7.50 shows the time courses of the UHF probes used to detect the place of origin of partial discharges.

In the practical application of this method, it is necessary to create a three-dimensional model of the transformer and formulate the influence of the environment on wave propagation, i.e., to take into account the paths

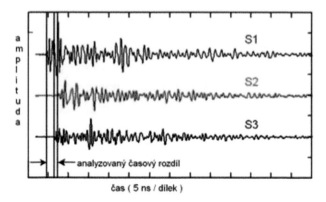

Figure 7.50   Time courses of output voltages of UHF probes.

of the wave in space and their possible reflections. Electromagnetic waves can also negatively affect very strong magnetic fields, usually acting in high-voltage devices. UHF signal sensing can also affect interference in the environment of power plants or HV substations.

7.2.3 Optical partial discharges detection

Optical detection of partial discharges is based on the detection of the electromagnetic field in the *infrared frequency spectrum* from 1 THz to 10^3 THz, in the *visible spectrum* in the range of 480–700 THz, and in the *ultraviolet spectrum* in the range of 10^3–10^5 THz.

Infrared radiation sensors allow you to measure the temperature or temperature pattern in the vicinity of the occurrence of partial discharges. In the visible spectrum of radiation, the nature of the discharges can be observed with the naked eye or recorded by optical instruments, such as cameras. To detect partial discharges in the *ultraviolet spectrum*, special ultraviolet cameras are used.

In all these cases, color *charge coupled device* (CCD) sensors are used to detect discharges, and their output image signals are processed by digital image processing methods, such as *contouring shapes*.

A typical commercial system for optical detection of discharges is the system of the Israeli company OFILSYSTEMS and its DayCor® camera, which operates in both ultraviolet and visible spectra. The block diagram of this camera is shown in Figure 7.51.

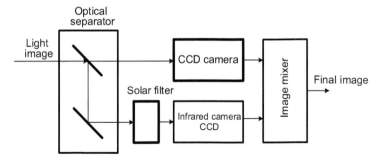

Figure 7.51 DayCor® Block Diagram.

To counteract the influence of external lighting on the quality of the image being scanned, the scanned light pattern is obtained by mixing an unfiltered image signal with a filtered signal using an ultraviolet filter.

This system can work in twilight, darkness, and in full daylight. The device is able to detect partial discharges with a charge of 1.3 pC at a distance of up to 10 m.

Figure 7.52 shows the use of the DayCor® camera in sensing partial discharges on HV and VHV distribution systems.

7.2.4 Chemical detection of partial discharges

The presence of discharge activity in oil- or gas-insulated equipment can also be detected by *chemical analysis of gaseous hydrocarbons* resulting from the discharge activity. These products accumulate in the insulator over the entire period of operation, and, therefore, according to their content, their degradation, which was caused by partial discharges, can be estimated. One of the most widely used methods of chemical gas analysis is gas chromatography *dissolved gas analysis* (DGA), which is used mainly in condition monitoring of HV and VHV transformers. This method is based on the analysis of the presence of hydrogen H_2, methane CH_4, ethylene C_2H_4, acetylene C_2H_2, and propylene C_2H_6 under the action of partial discharges in their solid and liquid insulators.

In this method, the analyzed gas is mixed with the so-called *carrier gas*, which is usually nitrogen N, helium He, or argon Ar. The resulting mixture is fed into a chromatographic column in which absorption (*GSC technique*) or dissolution (*GLC technique*) splits the mixture of gases between the stationary and mobile phases. The interaction of the two components of the gas mixture leads to their retention and their washing out (elution).

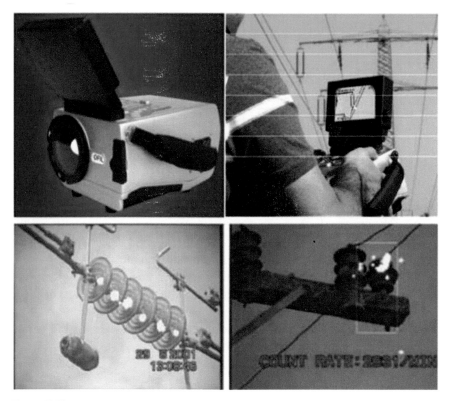

Figure 7.52   Example of using DayCor® to detect partial discharges on HV and VHV distribution systems.

The resulting components of the gas mixture are most often analyzed by the *flame ionization detector* (FID), in which the detection of gases is based on the formation of both positive and negative ions during the combustion of the analyzed substance in a hydrogen flame. The resulting ions of gases, which are carriers of charge, cause a current in the electrostatic field in the range of tens of pA to tens of nA. This current is amplified to tens to hundreds of megavolts in the electrometric amplifier. The block diagram of the gas chromatograph including the flame detector is shown in Figure 7.53.

The result of this analysis is a chromatogram that shows the time dependence of the voltage pulses, and their time remoteness is determined by the type of analyzed components of gaseous hydrocarbons C_2H_2, C_2H_4, CH_4, and C_2H_6, and whose area determines their concentration. Based on the concentration ratios of these gases, the intensity of partial discharges can then be approximately determined (Table 7.1).

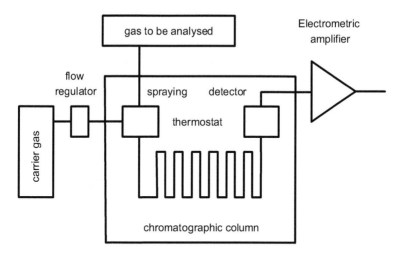

a) chromatograph block diagram

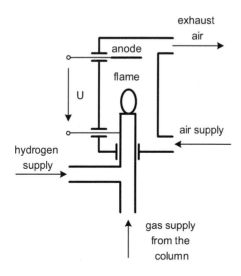

b) flame detector

Figure 7.53   Gas chromatograph.

Table 7.1   Determination of the intensity of partial discharges from the concentration ratio.

Intensity of partial discharges	C_2H_2/C_2H_4	CH_4/H_2	C_2H_6/C_2H_2	C_2H_2/CH_4
Low	< 0.1	< 0.1	> 0.5	< 0.3
Medium	≈ 0.4	≈ 0.3	≈ 0.3	≈ 0.3
High	> 0.7	> 0.5	< 0.1	> 0.3

References

[1] IEC 60 270 *Partial discharge measurement.* IEC 2000

[2] Florkowski, M. *Partial Discharges in High-voltage Insulating.* Wydawnictwa AGH 2020, ISBN: 97-8836636-4752

[3] Hauschild, W., Lemke, E. High Voltage Test and Measuring Techniques. Springer Verlag 2018. ISBN: 978-3-319-97459-0

[4] Cho, S. I. *On-Line PD (Partial Discharge) Monitoring of Power System Components.* Master's thesis, Aalto University, Finland 2011

[5] Boltze, M., Coenen, S., Tenbohlen, S. *Prospects and Limits of on-site PD Measurement.* TU Stuttgart, Germany 2011

[6] Haefely Test AG. *Breaking the limit of power capacitor resonance frequency with help of PD pulse spectrum to check and setup PD measurement.* hipotronics.com/document/breaking-the-limit-of-power-capacitor-resonance-frequency-with-help-of-pd-pulse-spectrum-to-check-and-setup-pd-measurement/

[7] Muhr, M. *Developments in Diagnosis of High Voltage Apparatus.* University of Technology Graz 2009

[8] Zalis, K. Evaluation of partial discharge measurememt. VII Symposium EUI 99. Zakopanie 21-23, Krakow 1999

[9] Kebbabi, L. *Optical and electrical characterization of creeping discharges over solid/liquid interfaces under lightning impulse voltage.* IEEE Transactions on Dielectrics and Electrical Insulation Vol. 13, No. 3, June 2006

[10] Lemke, E. etc. *Guide for Electrical Partial Discharge Measurements in compliance to IEC 60270*

[11] *Standard Test Method for Detection and Measurement of Partial Discharge (Corona) Pulses in Evaluation of Insulation Systems.* ASTM D1868

[12] *Partial Discharge Detection in Installed HV Extruded Cable Systems.* CIRGE Technical Brochure No. 182. WG 21.26, CIRGE, 2001

[13] *Guide for Electrical Partial Discharge Measurement in Compliance to IEC 60 270.* CIRGE Technical Brochure No. 366, CIRGE, 2008

[14] *UHF Partial Discharge Detection System for GIS.* Application Guide for Sensitivity Verification. Brochure 654, WG D1.25. CIRGE, 2016

[15] *Partial Discharges in Transformer.* Technical Brochure, No 676, WG D1.29, CIRGE, 2017

[16] *On-Site Partial Dicharge Assessment of HV and EHV Cable Stytems,* WG B1.28, CIRGE, 2018

[17] IEC TS 62478. *High Voltage Test Techniques – Measurement of Partial Dicharge s by Electomagnetic and Acoustic Methods*

[18] IEC 60 034-18-42.Rotating Electrical Maschienes – *Part 18 – 42. Partial Discharge Resistant Electrical Insultation Systems.*

[19] IEC TS 60034-27-2. *Rotating Electrical Maschienes. On-line Partial Discharge Measurement on the Stator Winding Insulation of Rotating Electrical Maschines.*

[20] IEC TS 61 934. *Electrical Insultating Materials and Systems - Partial Discharge Measurement.*

8

Voltage Tests of HV Electric Machines

These tests are defined for HV up to 50 kV by EN 60 052 and EN 60 060. When carrying out these tests, it is necessary to consider that some electric machines have non-renewable insulating materials, i.e., that after their penetration, their insulating properties are not restored to their original extent.

8.1 DC Voltage Tests

DC voltage tests shall be carried out at a voltage, the ripple of which shall be less than 3% of its nominal value. We recognize endurance tests, jump, and guaranteed jump voltage.

In the *durability voltage test*, the test voltage shall be continuously or gradually increased at a rate of approximately 2% of the nominal value of the machine voltage U_m per 1 s to 75% of this nominal value of the voltage at which it is maintained for 60 s and subsequently reduced to zero by discharging the charge of the capacity of the insulation system of the machine through the discharge resistance (Figure 8.1). The success of the exam is conditioned by the absence of a voltage jump.

In the *jump voltage test*, the test voltage is increased until a voltage jump (insulation breakthrough) occurs. Since the jump voltages have a considerable variance, a greater number of tests are performed at least 10 times, and the jump voltage is determined by their mean value. The machine under test satisfies this test if this means voltage value does not reach the guaranteed value of the machine's jump voltage.

8.2 AC Voltage Tests

AC voltage tests shall be carried out at alternating voltages with frequencies ranging from 45 to 65 Hz with a distortion of less than 5%. Similar

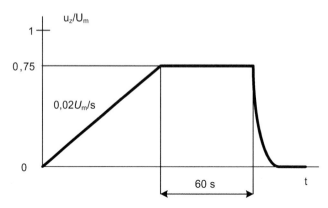

Figure 8.1   Time course of the voltage during the endurance voltage test.

to DC voltage tests, these tests are also tested with endurance, skip, and guaranteed jump voltages with the same procedure and method of their evaluation.

8.3 Voltage Tests by Long-term Induced Voltage

In this test, called *the step voltage test*, the test voltage is first increased at a rate of approximately 2% of the nominal voltage value of the U_m machine in 1 s. When $0.7 \cdot U_m$ is reached, this level of voltage is maintained for 1 minute. Then the voltage increases to $0.9 \cdot U_m$, on which it remains for 5 minutes. In the next phase of the test, the test voltage is first increased to the nominal value of U_m for 5 s and then reduced to $0.9 \cdot U_m$ for 30 minutes.

In the last phase of the test, the voltage is reduced to $0.7 \cdot U_m$ for 1 minute and then reduced to zero. This type of test is most often used for insulation tests of HV and VHV transformers. The time course of the induced test voltage is shown in Figure 8.2.

8.4 Impulse Voltage Tests

Atmospheric and switching voltage pulses are used for voltage tests of electrical HV devices. These voltage pulses test the puncture resistance of the insulation systems of electrical equipment. Atmospheric pulses simulate the time course of voltage in these devices during a lightning strike. Switching pulses correspond to the time course of the voltage during switching and switching processes. The time courses of both types of pulses are defined by EN 60 0601.

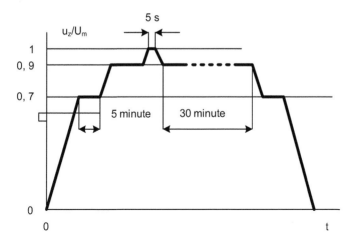

Figure 8.2 Time course of the test voltage during the long-term induced voltage test..

8.4.1 Course of impulse voltage tests

For these voltage tests, the following test procedures are generally used.

In the *first test procedure*, three pulses stress the insulation of the machine. The test is successful when the insulation does not break through. This test is recommended for non-renewing insulation.

In the *second test procedure* of the test, 15 pulses stress the insulation of the machine. The test is successful if there are no more than two punctures in the non-renewing insulation.

In the *third test procedure*, three pulses stress the insulation of the machine. The test is successful if there is no skipping. If the renewing insulation undergoes one puncture, another nine pulses are generated, and if even then the insulation does not break, the test is successful.

In the *fourth test procedure*, a larger number of pulses is generated, e.g., 100. In the case of renewed insulation, the stresses U_{10} at which there is a 10% probability of insulation breakthrough and the voltage U_{50} at which there is a 50% probability of insulation breakthrough are determined. If the voltage is U_{10}, which is less than the withstand test voltage, then the insulation of the machine passes the test.

8.4.2 Test pulses

The atmospheric pulse is defined by the time of the forehead and the time of the half-rear. The *forehead time T_c* is defined by 1.67 times the time of the interval T_0 between points A and B, at which the voltage pulse of 0.3 and 0.9

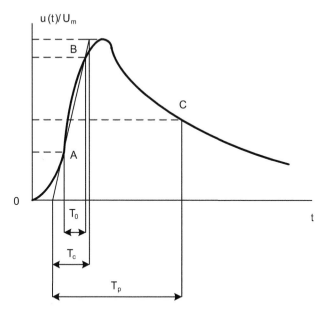

Figure 8.3   Atmospheric pulse time course.

takes on its amplitude U_m, respectively. The *half-rear time* T_p is defined by the time specified by the times when the line guided by points A and B intersects the timeline and the time when the voltage pulse reaches half the value of its amplitude when it decreases (Figure 8.3).

The *standard atmospheric pulse*, referred to as the 1.2/50 pulse, has a forehead time $T_f = 1.2$ μs with a tolerance of 30% and a half-rear time $T_{hr} = 50$ μs with a tolerance of 20%.

If, during the atmospheric pulse test, the voltage jumps or the insulation breaks, the time course of the atmospheric pulse is interrupted either in the forehead or in the rear.

In the case of a break in the forehead, the duration of the T_2 pulse is determined between the moments of intersection of the line, guided by points A and B with the timeline and the moment when the pulse is terminated. In relation to the duration of this pulse, the *collapse time* T_z is defined, which is 1.67 times the TCD time interval between points C and D, at which the pulse takes on 0.1 and 0.7 times its amplitude (Figure 8.4a), respectively. Similarly, the time of collapse when the time course of the atmospheric pulse in the rear is interrupted is defined (Figure 8.4b).

According to EN 60 0601, the switching pulse is defined by the *peak time* T_p at which the pulse reaches its amplitude and the *half-rear time* T_h, when its voltage drops to 1/2 of its amplitude (Figure 8.5).

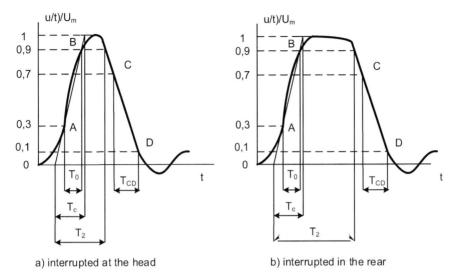

Figure 8.4 Time courses of interrupted atmospheric pulses.

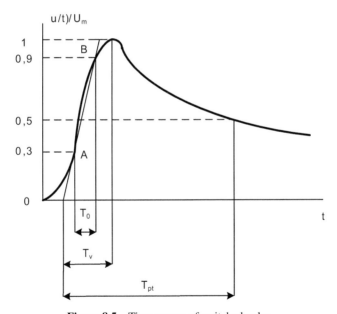

Figure 8.5 Time course of switched pulse.

A typical switching pulse has a peak time $T_p = 250$ μs with a tolerance of 20% and a half-rear time $T_{hr} = 2.5$ μs with a tolerance of 60%.

Table 8.1 shows the maximum permissible working, testing, induced, and surge voltages of electrical machines depending on their rated power.

Table 8.1   Test voltage of electric machines.

Nominal voltage stroke [kV]	Maximum working voltage [kV]	Test voltage [kV]	Induced voltage [kV]	Surge voltage [kV]
1	1.1	3	2	–
6	7.2	22	12	60
10	12	28	20	75
22	25	50	44	150
35	38.5	75	70	190
110	123	230	220	550
220	245	460	440	1050

8.4.3 Generation of test voltage pulses

Single-stage shock generators are used for the generation of test pulses up to amplitudes of tens of kilovolts, and *multi-stage shock generators* are used for the generation of pulses with amplitudes up to MV units.

8.4.3.1 Single-stage voltage shock generators

They consist of a capacitor C_1, which is charged from a DC voltage source via the R_{nab} charging resistance. After charging the capacitor, the charging circuit is disconnected, and the generation of the voltage pulse is started by jumping the voltage on the spark gap J. This charges capacitor C_2 through resistor R_1, while the time constant $R_1 \cdot C_2$ determines the time of the front of the generated pulse. The time of the pulse is determined by the time constant $R_2 \cdot C_2$ (Figure 8.6).

Based on circumferential differential equations,

$$u_1(t) = R_1 C_2 \frac{du_2(t)}{dt} + u_2(t), \qquad -C_1 \frac{du_1(t)}{dt} = \frac{u_1(t)}{R_2} + C_2 \frac{du_2(t)}{dt} \qquad \text{(8.1a, b)}$$

can be expressed as the time change of the generated pulse on the capacitor C_2

$$\frac{du_2(t)}{dt} = \frac{u_1(t)}{R_2 . C_1} - \frac{u_1(t) + u_2(t)}{R_1 . C_1}. \qquad (8.2)$$

The proportion of the output and input capacities of the generator and the ratio of the square of the amplitude ratio of the generated and charging voltages determine the efficiency of the generator, reaching 0.85–0.95

$$\eta = \frac{C_2}{C_1} \left(\frac{U_m}{U_n} \right)^2. \qquad (8.3)$$

Single-stage generators achieve an efficiency of up to 85%.

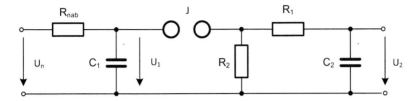

Figure 8.6 Single-stage voltage shock generator.

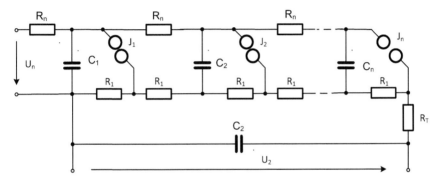

Figure 8.7 Marx's multistage voltage shock generator.

8.4.3.2 Multi-stage voltage shock generators

They are formed by a cascade of HV capacitors C_i, i = 1, 2,..., n, which are charged via R_n rechargeable resistors and spark gaps $J_1–J_n$ (Figure 8.7).

Their charging time must be so long that all capacitors are charged to charges $Q = U_n \cdot C_i$. When they are charged, there is the same voltage between the electrodes of all spark gaps, with one spark gap electrode grounded and the other connected to one of the electrodes of the charged subsequent capacitor.

The first, the so-called trigger spark gap J_1, contains an additional electrode to which a trigger pulse is applied, the amplitude of which is 5%–20% higher than the voltage on the capacitor C_1. This will jump the voltage on this spark gap, causing a serial connection of capacitors C_1 and C_2, which then have twice the charging voltage relative to the charging voltage.

This voltage causes another jump in the next spark gap J_2, resulting in a serial connection of the capacitor C_3 already to the series-connected capacitors C_1 and C_2. At the end of this chain reaction, after the voltage jump at the last spark gap J_n, the load capacitor C_2 is charged to the voltage $n \cdot U_n$, where n is the number of stages of the generator.

On the shape of the occipital part of the pulse have the main share resistors R_2, whose resistances are several times larger than those of resistors R_1.

Table 8.2   Magnitude of the constants k_1 and k_2 for atmospheric and switching pulses.

Pulse type	Atmospheric	Switching
k_1	2.06	2.41
k_2	0.73	0.87

The additional resistor R_p dampens oscillations arising from the parasitic inductance of the L_p external pins between the generator and the load.

The time course of the generated pulses is expressed by the function

$$U_2(t) = \frac{U_n}{R_1 C_2} \frac{\tau_1 \tau_2}{\tau_2 - \tau_1} \left(e^{-\frac{t}{\tau_2}} - e^{-\frac{t}{\tau_1}} \right), \tag{8.4}$$

where

$$\tau_1 = R_1 \frac{C_1 . C_2}{C_1 + C_2}, \qquad \tau_2 = R_2 (C_1 + C_2). \tag{8.5a, b}$$

are time constants determining the shape of the generated pulses.

From the previous equations, it is possible to determine the resistors, determining the time of the forehead and the time of the half of the individual generated pulses

$$R_1 = \frac{T_c}{k_1} \frac{C_1 + C_2}{C_1 C_2}, \qquad R_2 = \frac{T_p}{k_2} \frac{1}{C_1 + C_2}. \tag{8.6a, b}$$

Table 8.2 shows the constants k_1 and k_2 for atmospheric and switching pulses.

Parasitic inductance L_p of external pins between the generator and the load causes undesirable oscillations at the peak of the generated pulses, which are dampened by the resistor R_d, for which the resistance must apply

$$R_d > 2\sqrt{L_p \frac{C_1 + C_2}{C_1 C_2}}. \tag{8.7}$$

For both types of generators, *spherical spark gaps* consisting of two ball electrodes are used (Figure 8.8).

The jump voltages of spherical spark gaps are given in Table 8.3 within the meaning of EN 60 052 under normal atmospheric conditions, 1013 hPa, temperature of density of 8.5 g/m³.

Since the electrical strength of air is influenced by its temperature, pressure, and humidity, it is necessary to correct the indicated values of jump voltages. For the temperature dependence of air density,

$$\delta = \frac{p}{p_0} \frac{273 + \vartheta_0}{272 + \vartheta}, \tag{8.8}$$

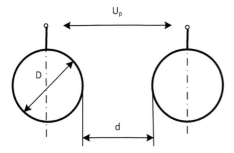

Figure 8.8 Spherical spark gap.

Table 8.3 Dependence of the jump voltage in kilovolts on the diameter D and their distances d.

D (cm)	d (cm)										
	0.5	1	2	3	4	5	6	7	8	9	10
2	17.4	30.7									
10	16.8	31.7	59	84	105	123	138	150			
50			59	86	112	138	164	189	214	239	263

where ϑ is the air temperature in °C, $p_0 = 1013$ hPa is its normal pressure, and p is its current pressure. Correction for air humidity is expressed by the following equation:

$$k_v = 1 + 0.002\left(\frac{v}{\delta} - 0.85\right).$$ (8.9)

The relation then determines the corrected jump voltage

$$U_{pk} = \delta U_{pn}.$$ (8.10)

where U_{pn} is the nominal value of the jump voltage under normal atmospheric conditions.

Figure 8.9 shows a multi-stage voltage shock generator with a nominal voltage of 110 kV with a measured transformer.

8.4.4 Voltage level impulse tests

In accordance with EN 60060-1, these tests are divided into three classes. The first class includes tests with multiple test voltage levels. The second class of tests defines the up–down tests. The third group of tests includes tests by gradual stress. For all these tests, statistical methods of processing the measured data are used to evaluate them.

Figure 8.9 Multistage voltage shock generator with metered transformer.

8.4.4.1 Tests with a constant number of voltage levels

These tests are usually used to test the electrical strength of HV insulation systems of electrical machines by pulse signals.

Before the actual measurement, it is necessary to determine the number of supplied voltage pulses n, through which the test at m voltage levels will be carried out. The measurement starts at such a voltage level at which there is no jump in isolation. Then the voltage levels are increased after the same values, evaluating the number of k jumps for each voltage level.

Figure 8.10 shows the time sequence of test pulses on the increasing attached voltage using six voltage levels with eight pulses at each of these voltage levels.

From the number of occurrences of k_i punctures at each of the levels, their probability of occurrence is determined

$$p_i = \frac{k_i}{n}, \quad i = 1, 2, ..., h, \tag{8.11}$$

where n is the number of all pulses at each of the levels and h is the number of levels.

8.4.4.2 Up–down tests

These tests are used to diagnose *regenerative insulators*. In particular, the tests are suitable for determining the magnitude of voltage levels at which there is a 10% and 90% probability of jumping in isolation. A maximum of seven

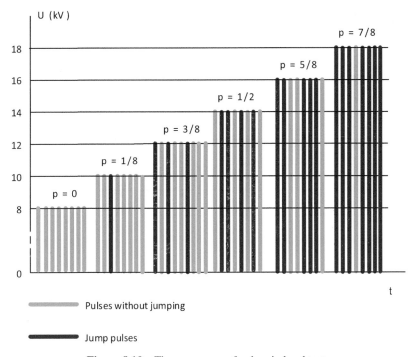

Figure 8.10 Time sequence of pulses in level test.

pulses are applied for each voltage level. Impulse levels will then decrease or increase depending on whether voltage levels of $U(10)$ or $U(90)$ corresponding to 10% or 90% of the probability of jumps are evaluated. Accordingly, we distinguish between *endurance* and *discharge test procedure*.

The *endurance* type *test* of the test is carried out in such a way that the voltage increases from the first selected voltage level. If none of the seven attempts is jumped, we increase the voltage by ΔU. When the jump occurs, we do not test the remaining experiments, but we reduce the voltage by ΔU and measure at a lower voltage level. The magnitude of the test voltage change should be in the range of $0.01 \cdot U_{50}$ to $0.03 \cdot U_{50}$. Each voltage level received is considered to be one at which at least two groups of pulses have been measured.

According to Figure 8.11, a voltage level of 16 kV is not considered acceptable.

The discharge type of the test gradually increases the voltage levels until there is no jump in the insulation. If, when the voltage levels increase, there is a jump at all pulses, then their voltage level is reduced to a lower value.

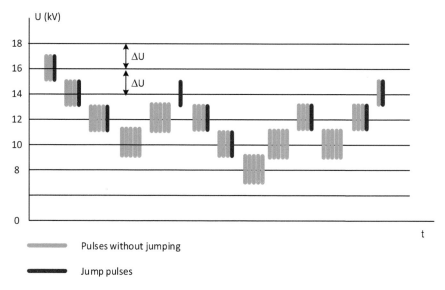

Figure 8.11   Time sequence of pulses in endurance test up down.

Figure 8.12 shows the time sequence of pulses during the up–down discharge test.

The voltage levels $U(10)$ and $U(90)$ are determined from the following equation:

$$U = \sum_{i=1}^{l} \frac{k_i U_i}{m},$$

where k_i is the number of pulse groups received for a given voltage level, i is the number of voltage levels received, U_i is the voltage of each of the voltage levels, and m is the total number of pulse groups received at all voltage levels.

8.4.4.3 Phased stress tests

In this test, the DC and AC voltages are increased until the voltage jump occurs in the insulation. When using impulse voltage waveforms, their amplitude is gradually increased until the first voltage jump occurs in isolation. After n tests performed, we get m of jump voltage values. The jump voltages at which the jump occurred are divided into groups according to their size.

Then, similarly to the level test, the probability of the occurrence of the number of jumps in each set of pulses is determined, from which the probability function of the jump voltages can be determined.

Figure 8.13 shows the time sequence of pulses during the successive stress test.

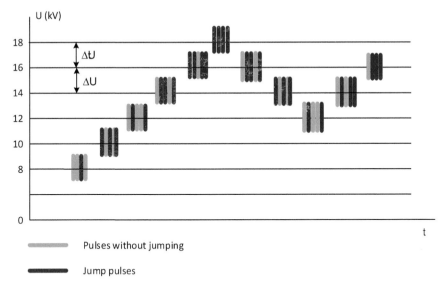

Figure 8.12 Time sequence of pulses in the up-down discharge test.

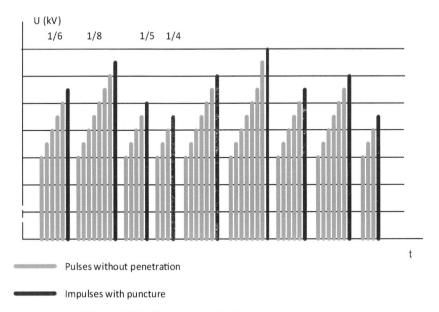

Figure 8.13 Time course of pulses under gradual stress.

8.4.4.4 Statistical methods for the evaluation of level tests

Since the discharge activity in insulators is always random in nature, statistical methods of data processing are used to evaluate it.

The basic statistical parameters include arithmetic mean, median, and standard deviation.

The *arithmetic mean* determines the most likely value from the measured data. It is defined by the simple sum of the samples of the set X_i, $i = 1$, $2, \ldots, n$, where n is their number

$$\mu = \frac{1}{n} \sum_{i=1}^{n} x_i. \tag{8.12}$$

Since a random variable X is determined by the distribution of probabilities of occurrence of p of its values, its *distribution function* $F(x_i)$ can be defined by the probability that this variable will acquire a value less than or equal to the value of x_i

$$F(x_i) = p(X < x_i). \tag{8.13}$$

A *quantile* is defined as the value of a random variable that divides data into two parts with higher and lower values than a specified quantity size. The most widely used quantile is the quantile $x(0.5)$, which divides the statistical set into two parts with identical probabilities. This quantile is called the median.

Standard deviation is used to assess the variance of the measured values

$$\sigma = \sqrt{\frac{1}{n-1} \sum_{i=1}^{n} (x_i - \mu)^2}. \tag{8.14}$$

To determine the properties of the frequency of jump occurrences in the insulation systems of HV electrical machines, it is advisable to use *binomial distribution*, since it is the occurrence of a random phenomenon in n independent experiments, the number of which is limited to several dozen.

The probability that the jump occurs k times in n independent experiments is determined by the following equation:

$$p(x,k) = \binom{n}{k} p^x (1-p)^{n-x}, \tag{8.15}$$

where

$$\binom{n}{k} = \frac{n!}{k!(n-k)!}, \tag{8.16}$$

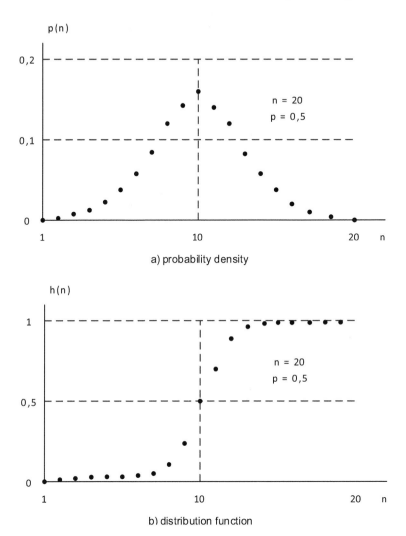

Figure 8.14   Binomial distribution.

and $n!$ and $k!$ are the factorials of n and k, which are the products of all integers less than n and k with the numbers n and k.

Figure 8.14 shows the *probability density* and its corresponding distribution function of the binomial distribution with the number of samples $n = 20$ and the probability of occurrence of samples $p = 0.5$.

The shape of this distribution depends on the number of attempts n. For small numbers of samples, its probability density is distributed over the entire interval. Conversely, for a large number of experiments performed, the probability density characteristic is more symmetrical and narrower, thereby increasing the accuracy of the estimate.

For the *probability density* of this distribution,

$$h_n = \frac{k}{n}.$$ (8.17)

The probability density is always less than or equal to 1. The remaining part, belonging to 1, is the relative frequency of endurance tests and can be determined by a complementary relationship

$$h_{nc} = 1 - h_n = \frac{n-k}{n}.$$ (8.18)

It follows that with an increasing number of n measurements for one voltage level, greater determination accuracy can be achieved. In the limit case, when the number of measurements is close to infinity, the probability of occurrence of the phenomenon approaches 1, i.e., that it is close to certainty.

In practice, however, we do not achieve a large number of processed samples, and, therefore, it is necessary to place the expected value of probability in an interval with a certain level.

For example, for the relative frequency $h_n = 0.7$ and the number of samples $n = 10$, the higher probability limit $p_h = 0.91$ and the lower limit $p_l = 0.37$.

It can be proved that the binomial distribution with several dozen samples almost coincides in its properties with the *normal (Gaussian) distribution* with probability density

$$P(x) = \frac{1}{\sigma\sqrt{2\pi}} e^{-\frac{(x-\mu)^2}{2\sigma^2}},$$ (8.19)

where μ is the arithmetic mean and σ is the standard deviation of the normal distribution.

For the distribution function of the normal layout,

$$F(x) = \frac{1}{\sigma\sqrt{2\pi}} \int_{-}^{x} e^{-\frac{(x-\mu)^2}{2\sigma^2}} dx.$$ (8.20)

In addition to the binomial and normal distribution, the *Gumbel* and *Weibull distributions* are used to evaluate voltage tests. The *Gumbel distribution* is suitable for evaluating tests of insulation with low electrical strength, where there are a high number of punctures and jumps. On the contrary, the *Weibull distribution* is suitable for analyzing the properties of high-electrical-strength insulators with a low number of jumps and punctures.

To compare the test results with the use of the above layouts, the *Kolmogorov–Smirnov good throw test* is most often used. This test determines whether the distribution of the measured data matches the selected distribution, which is most often a binomial or normal distribution.

An ordered random data sample $(x_1, ..., x_n)$ creates a *distribution selection function* $F_n(x)$

$$F_n(x) = 0, \text{pro } x < x_1, \tag{8.21a}$$

$$F_n(x) = \frac{i}{n}, \text{pro } x_i < x < x_{i+1}, i = 1, 2, ..., n-1, \tag{8.21b}$$

$$F_n(x) = 1, \text{pro } x > x_n. \tag{8.21c}$$

This selective distribution function is compared with the distribution function of the selected distribution $F(x)$. The test criterion is the so-called *Kolmogor statistics* D_n, the value of which is determined by the maximum absolute difference between the sample distribution function $F_n(x)$ and the distribution function of the chosen distribution $F(x)$.

These values are compared with the critical values that are given in the relevant statistical tables for the different significance levels, which represent the probability with which the hypothesis can be erroneously rejected.

8.5 Shock Wave Tests

Shock wave tests are *impulse-winding tests* of all electrical machines. These tests are carried out not only in the manufacture of electric machines but also during their use in practice. The tests make it possible to detect *inter-threaded shortfalls* between individual windings and a different number of turns between individual phase windings.

In these tests, the windings of the machines are stressed by current pulses with very short rise times, which are generated by discharging capacitors C_1 and C_2 into the diagnosed windings. The amplitude of these pulses is usually selected in the range of twice the nominal voltage of the windings, increased in addition by 1 kV.

Figure 8.15 shows the principle scheme of a simple impact generator.

HV voltage waveforms on diagnosed windings are displayed by a digital oscilloscope after their amplitude reduction in the voltage dividers or recorded by a notebook of transient processes. Assuming that one of the coils does not have a malfunction, the time course of the time response of the second diagnosed coil can be inferred to the type and extent of failures in its winding.

Figure 8.16 shows examples of time responses of both failure-free coils of the HV transformer and the time responses of the coils of this transformer, of which there is an inter-threaded short circuit in the second coil.

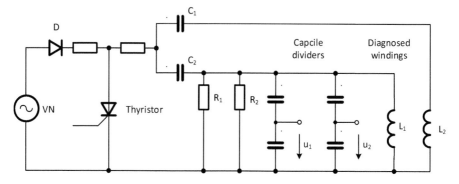

Figure 8.15   Principle diagram of the dimple impact shock generator.

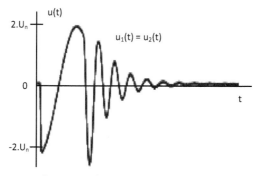

a) trouble-free state of windings

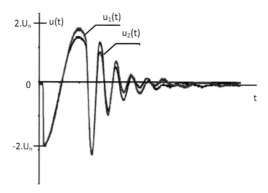

b) inter-threaded short circuit in the winding

Figure 8.16   Time courses of voltage on HV transformer coils.

The rate of difference between the time responses of coils is usually expressed either by the *difference in the amplitudes* of these responses relative to the amplitude of the response of the trouble-free winding

$$k = \frac{U_{1m} - U_{2m}}{U_{1m}},$$ (8.22)

or the difference of the surfaces below these responses, related to the response area of the trouble-free winding

$$l = \frac{\sum_{=1}^{n} |u_{i1} - u_{i2}|}{\sum_{=1}^{n} |u_{i1}|}.$$ (8.23)

where u_{i1} and u_{i2} are voltage response samples $u_1(t)$ and $u_2(t)$.

References

[1] Hauschild, W., Lemke, E. High Voltage Test and Measuring Techniques. Springer Verlag 2018. ISBN: 978-3-319-97459-0.

[2] Hauschild, W. *High-voltage test and measuring techniques*. New York: Springer, 2013, ISBN: 978-3642-45-3519.

[3] Kuffel, E, W Zaengl a J Kuffel. *High voltage engineering: fundamentals*. 2nd ed. Boston: Butterworth-Heinemann, 2000. ISBN: 97-8-0750-63634-6

[4] Malik, N, A., Qureshi, M. *Electrical insulation in power systems*. New York: Marcel Dekker, 1998. ISBN 082-470-106-2

[5] Guastavino, F., A. Ratto, F. Porcile, E. Torello A D. Santinelli. *Dielectric characterization of gas mixtures as electrical insulating for high voltagecomponents and appliances*. 2013 Annual Report Conference on Electrical Insulationand Dielectric Phenomena. IEEE, 2013, ISSN: 1008-1011

[6] *Electric Field Intensity. High Voltage Engineering*. http://nptel.ac.in/courses/108104048/ui/Course_home1_3.htm

[7] Schwenk, K., Wolf, J. *SGS Impulse Voltage Generator* - Manual. 2006.

[8] Haefely Hipotronics. http://www.haefely-hipotronics.com/.

[9] Weedy, B. M. *Electric Power Systems*, Wiley 2012, ISBN: 978-0-4706-8268-5

[10] SSTM D1869. *Standard Test Method for Detection and Measurement of Partial Discharge Pulses in Evalution of Insulation Systems.*

[11] CIRGE Technical Brochure No. 703. *Insuklation Degradation Under Fast, Repetitive Voltage Pulses.* WG D1.43, 2017

[12] IEC 600 0347. *Rotating electrical machines - Part 7: Classification of types of construction, mounting arrangements and terminal box position* (IM Code), 2020.

9

Diagnostics of Power Transformers

High-voltage transformers play an irreplaceable role in the distribution and transmission of electricity in the electricity system.

9.1 Types of Power Transformers

According to the method of use, we distinguish distribution and block transformers.

Distribution transformers are used to transmit energy directly to end consumers. The voltage levels of the primary windings of distribution transformers correspond to HV voltage levels of 3, 6, 10, 22, and 35 kV. On the secondary side, there is always a voltage level of 400 V. Power of distribution transformers does not exceed several mega volt amps (MVA).

Distribution transformers transform the voltage levels of generators, or other power generation points, into the transmission and distribution electrical system. The voltage levels of the primary windings of distribution transformers correspond to the HV (*high voltage*) levels. The voltage levels of the secondary windings of power transformers correspond to VHV (*very high voltage*) levels of 110 and 220 kV and to UHV (*ultra-high voltage*) levels of 400 and 750 kV.

According to the transmitted power, transformers are divided into low-power transformers in the range of 500–7500 kVA, medium-power transformers in the range of 7.5–100 MVA, and transformers with a high power greater than 100 MVA.

Special power transformers are transformers used in HVDC (*high voltage direct current*) power transmission systems. These transformers are designed with minimal losses even at higher frequencies, because the produced AC current is distributed to the DC transmission network after transformation and rectification. Compared to the AC distribution network, there

a) distribution transformer   b) distribution transformer

Figure 9.1   Typical types of transformers.

are no reactive losses in the DC network, which are caused by the AC capacities and inductances in the AC networks. At present, there are DC networks
in the world with nominal DC voltages of 100, 400, and 850 kV.

From the point of view of *transformer cooling*, we distinguish between
transformers, so-called dry compact transformers, cooled only by air, and
transformers cooled by oil filling, or with additional air cooling.

Figure 9.1 shows a typical three-phase compact distribution transformer
22 kV/400 V with a power of 10 MVA and a distribution transformer 22 kV/
110 kV with a power of 250 MVA with oil cooling, supported by air cooling.

9.2 Properties of Transformers

The principle of operation of transformers is based on the *Maxwell–Faraday
induction law*, which defines the induced voltage $u_i(t)$ by the number of turns
N, in which this voltage is induced due to the time change of the induction
$d\Phi/dt$ in its ferromagnetic core

$$u_i(t) = -N \frac{d\Phi}{dt}. \tag{9.1}$$

If we apply an alternating voltage, U_1, to the primary winding of the transformer, then the current i_1 passing through the primary winding of the transformer with N_1 turns generates an alternating magnetic flux in the magnetic
circuit of the transformers, which induces an alternating voltage in the secondary winding N_2 of the *unloaded transformer*.

$$U_{i2} = \frac{N_2}{N_1} U_1. \tag{9.2}$$

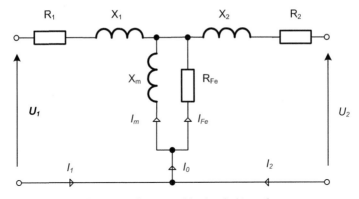

a) replacement diagram of the loaded transformer

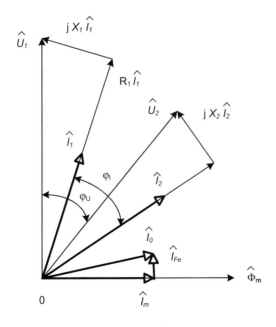

b) phasor diagram

Figure 9.2 Replacement diagram of a loaded transformer.

The ratio $p = U_2/U_1 = N_2/N_1$ defines the *nominal conversion* of the transformer.

The properties of the actual loaded transformer are expressed by its substitute diagram and the corresponding phasor diagram, which shows the phasors of the primary and secondary voltages and currents during its unit conversion (Figure 9.2).

In this surrogate circuit, resistors R_1 and R_2 and reactance $X_1 = \omega \cdot L_1$, $X_2 = \omega \cdot L_2$ represent the parameters of the primary and secondary windings of the transformer. The reactance X_m expresses the scattering of the magnetic flux outside the transformer core and the resistor R_{Fe} expresses the losses in the transformer's magnetic circuit. The phase shifts φ_U and φ_I determine the phase shifts between the primary and secondary voltages and between the primary and secondary currents of the transformer.

The extreme activities of each transformer include its *no-load opera-tion at* zero output current I_2 and short-circuit operation at its short-circuited output, i.e., at $U_2 = 0$.

At *no-load*, when the current of the transformer's secondary winding is zero, the current in the transformer's primary winding is determined by the magnetizing current I_m, which is caused by eddy currents and losses in its ferromagnetic core.

Eddy currents are created by magnetic induction of current in the con-ductive ferromagnetic core of the transformer and are inversely proportional to its magnetic resistance. The elimination of these currents is achieved by insulating the transformer sheets, of which the magnetic circuit of the trans-former is composed. Eddy current losses are proportional to the squares of frequency f, magnetic induction B, and transformer plate thickness t

$$P_v = k_v \cdot f^2 \cdot B^2 \cdot t^2. \tag{9.3}$$

Hysteresis losses are caused by overmagnetization of the transformer's fer-romagnetic core due to its excitation by alternating current. These losses are proportional to the frequency of the excitation current f and the square of the magnetic induction B

$$P_h = k_h \cdot f \cdot B^2. \tag{9.4}$$

During *short-circuit operation*, when the secondary winding of the trans-former is short-circuited, the current in the primary winding determines the losses caused by its resistances in the primary and secondary windings. In this connection, the *short-circuit voltage U_{1k}* is defined as the voltage on the primary winding of the transformer at its rated primary current.

The *losses caused by the winding resistances* are determined by the sum of the squares of the flowing currents I_i and the winding resistances R_i in all n transformer windings.

$$P_{Cu} = \sum_{i=1}^{n} R_i \cdot I_i^2. \tag{9.5}$$

During short-circuit operation, there is an increase in scattering magnetic fluxes, which, in the ferromagnetic core of the transformer and in its windings, cause additional losses ΔP_d, which are similar in nature to eddy currents.

The *efficiency of the transformer* is given by the ratio of its output power P_2 to the input power P_1

$$\eta = \frac{P_2}{P_1} = \frac{P_1 - \Delta P}{P_1}, \qquad (9.6)$$

where $\Delta P = \Delta P_0 + \Delta P_k$ are its *total losses*, given by no-load losses ΔP_0 and short-circuit losses ΔP_k.

The analysis mentioned so far, which applies to *single-phase transformers*, also applies to an extended extent to *three-phase transformers*, with the difference that the electrical parameters are assessed not only in the individual phases of the transformer but also the entire three-phase transformer.

The efficiency of power transformers increases with their rated power. For example, transformers with an output of hundreds of kilovolt amperes achieve efficiencies of up to 95%, and transformers with outputs of several tens to hundreds of mega volt amps achieve an efficiency of up to 98%.

Various types of mineral or synthetic transformer oils are usually used to cool power transformers, which also act as insulators. The increase in cooling efficiency of high-power transformers is achieved by their additional forced air cooling.

9.3 Diagnostics of Electrical Properties of Power Transformers

The basic diagnostic electrical quantities used in the non-disassembly diagnostics of MV transformers include verification of their conversion ratios, determination of phase sequence and clock angle, winding resistances, short-circuit voltage and current, leakage current and insulation resistance, capacity, loss and polarization factor, and ratio capacitances at 2 and 50 Hz. Low voltage pulse methods and winding impedance frequency response methods are used to detect winding displacement.

Other tests of transformers include dielectric tests (IEC 600076-3), tests with applied alternating voltage, endurance induced voltage tests, and measurement of partial discharges (EN 60270). Transformer-type tests include warming test (IEC 600076-2), dielectric test (IEC 600076-3), and atmospheric pulse test.

Special tests of transformers include dielectric special tests (IEC 600076-3), voltage-induced voltage test with partial discharge measurement, atmospheric pulse wave test, atmospheric impulse test at terminal N, test of atmospheric overvoltage pulses applied simultaneously to multi-phase terminals, short-circuit resistance test (IEC 60076-5), noise level determination (IEC 60076-10), environment class test, climatic test, thermal shock test, and flammability test.

9.3.1 Measurement of active winding resistance

The active resistance of the winding is measured at all windings of the disconnected transformer; while measuring one of the windings, the other windings are disconnected. The magnitude of the current flowing through the measured winding should not exceed 20% of the rated current of this winding. The minimum current value should be at least , where I_{10} is the RMS value of the no-load current at the nominal primary voltage of the transformer. It is also necessary to measure on a temperature-stable transformer so that temperature changes in the winding resistance do not apply during the measurement.

If all the winding terminals of the three-phase transformer are accessible, then their resistances can be measured separately. If the windings are permanently *connected to the star*, then the resistances of two windings connected in series are measured (Figure 9.3a)

$$R_{UV} = R_U + R_v, \qquad R_{UW} = R_U + R_w \qquad R_{VW} = R_V + R_w. \qquad (9.7a, b, c)$$

If the windings are permanently *connected in a triangle*, then the resistance of one of the windings, which is connected in parallel with the remaining windings connected in series, is always measured (Figure 9.3b)

$$R_{UV} = \frac{R_U(R_V + R_W)}{R_U + R_V + R_W}, \qquad R_{UW} = \frac{R_W(R_U + R_V)}{R_U + R_V + R_W}, \qquad R_{VW} = \frac{R_V(R_U + R_W)}{R_U + R_V + R_W}.$$

$$(9.8a, b, c)$$

The winding resistances are similarly determined when they are *connected to a refracted star* (Figure 9.3c).

If the resistances of the individual winding sections are approximately identical, then from the measured resistances R_{UV}, R_{UW}, and R_{VV}, it is possible to determine the mean winding resistance of the respective winding phase

$$R_s = \frac{R_{UV} + R_{UW} + R_{VW}}{6}. \qquad (9.9)$$

If the measured resistances between the phases are *approximately the same*, then for the winding resistances of the individual phases, $R_f = 0.5 \cdot R_s$ for the

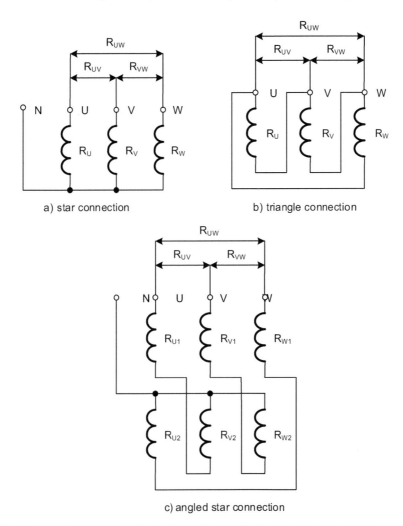

a) star connection   b) triangle connection

c) angled star connection

Figure 9.3 Measurement of the active resistance of a three-phase transformer winding.

star connection and $R_f = 1.5 \cdot R_s$ for the delta connection. Similarly, the resistances of the winding, the transformer, connected to the refracted star can be determined.

A *four-wire connection* method is used to measure the winding resistances, in which the measured resistance flows through a current from a current source and the magnitude of the measured resistance is determined from the voltage drop at the measured transformer terminals. This method eliminates the effect of lead resistances on measurement accuracy.

When measuring the resistance of the winding, it is necessary to take into account the *resistance temperature dependence*. The same applies to temperature corrections of copper and aluminum winding resistances

$$R_{2Cu} = R_{1Cu}\frac{235 + \vartheta_2}{235 + \vartheta_1}, \qquad R_{2Al} = R_{1Al}\frac{225 + \vartheta_2}{225 + \vartheta_1}, \qquad \text{(9.10a, b)}$$

where R_{1Cu} and R_{1Al} are winding resistances at temperature ϑ_1 and R_{2Cu} and R_{2Al} jsou are winding resistances at temperature ϑ_2

The actual inductance L of the transformer winding causes an exponential increase in current when the voltage U is connected to its winding

$$i(t) = \frac{U}{R}\left(1 - e^{\frac{t}{\tau}}\right), \qquad \text{(9.11)}$$

where $\tau = L/R$ is the *winding time constant*.

The time constants of HV transformers can reach up to tens of seconds.

If we require that the measurement error caused by the final current stabilization time is less than its entered value ε, then for the current stabilization time,

$$T_u = \tau.\ln\left(-\frac{1}{\varepsilon}\right). \qquad \text{(9.12)}$$

For example, for a measurement error of 1%, the settling time is 4.6·τ, for a error of 0.1% already 6.9·τ.

9.3.2 Checking the sequence and rotation of the phases

In order for the transformers to run in *parallel*, it is necessary to observe, in addition to the same nominal primary and secondary voltages of the transformers, their same sequence and phase rotation.

It is obvious that a cyclic change of the terminals of a three-phase transformer does not change its phase sequence, and, conversely, a non-cyclic change of its terminals always changes its phase sequence.

The *standard phase sequence* is considered to be the sequence in which the voltage phasor $\widehat{U}_U$ is delayed by 120°, with respect to the voltage phasor $\widehat{U}_V$ and the voltage phasor is delayed by 120°, with respect to the voltage phasor $\widehat{U}_W$.

To determine the phase sequence, phase sequence indicators were used, which were previously formed by a system of three excitation coils, geometrically rotated by 120°, whose rotating magnetic field induced an alternating electric current in the conductive disk, the force effects of which rotated

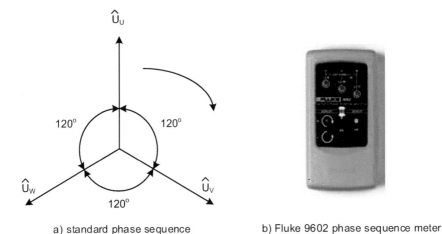

a) standard phase sequence   b) Fluke 9602 phase sequence meter

Figure 9.4   Phase sequence and its measurement.

the disk. The direction of rotation of the disk was determined by the phase sequence of the three-phase voltage system. Modern phase sequence indicators use phase-shifted currents to excite the LEDs to determine the phase sequence.

Figure 9.4 shows the standard phase sequence of a three-phase voltage system and a phase sequence meter with FLUKE 9062 LEDs.

In connection with the control of the phase sequence, the so-called clock angle is determined for transformers, which is given by the phase shift of the voltage between the equally marked input and output terminals of the transformer. The hour angle is given in multiples of a phase shift of 30°. The phase rotation of the voltage phasors on the transformer is indicated by the symbols Y, D, and Z, which express the method of connecting its winding with a higher voltage, and the symbols y, d, and z, which express the method of connecting its winding with a lower voltage. The letter Y, respectively y, denotes the connection of the winding to the star, the letter D, respectively d, denotes the connection of the winding to the triangle, and the letter Z, respectively z, indicates the connection of the winding to the angled star.

For example, the designation $Yd5$ corresponds to a higher voltage winding to a star, a lower voltage winding to a triangle with a corresponding clock angle of 5 hours, which is 150°. Twenty-six possible connections are allowed, the most used being Yy, Dy, Dz, Yd, and Yz.

The connection is recommended at approximately symmetrical load of the transformer, the connection Yz at its significantly asymmetrical load, the

Table 9.1 Determination of phase shift from voltage on higher and lower voltage winding sections.

$U_{Vv} = U_{Wv}$ $U_{Vw} = U_{Wv}$	$U_{Vv} < U_{Vw}$		$\varphi = 0$
	$U_{Vv} > U_{Vw}$		$\varphi = 180°$
$U_{Vv} = U_{Wv} = U_{Vw}$	$U_{Vv} < U_{Yw}$		$\varphi = 30°$
	$U_{Yv} > U_{Vw}$		$\varphi = 210°$
$U_{Vv} = U_{Vw} = U_{Ww}$	$U_{Vv} > U_{Wv}$		$\varphi = 150°$
	$U_{Vv} < U_{Wv}$		$\varphi = 330°$
$U_{Vv} = U_{Ww}$	$U_{Vv} < U_{UV}$ $U_{Vw} > U_{UV}$		$\varphi = 60°$
	$U_{Vv} < U_{UV}$ $U_{Vw} < U_{UV}$		$\varphi = 300°$
	$U_{Vv} > U_{UV}$ $U_{Vw} > U_{UV}$	$U_{Vv} > U_{UV} + U_{uv}$	$\varphi = 180°$
		$U_{Vv =} U_{UV} + U_{uv}$	$\varphi = 120°$
	$U_{Vv} > U_{UV}$ $U_{Vw} < U_{UV}$	$U_{Wy} = U_{UV} + U_{uv}$	$\varphi = 240°$
		$U_{Vw} < U_{UV} + U_{uv}$	$\varphi = 270°$

connection *Dy* at unbalanced load and high power, and *Yd* at extremely high load in HV networks.

Hour angles can also be measured using a voltmeter by connecting two identically marked terminals on the higher and lower voltage sides, e.g., terminals u and U and terminals with higher voltage supplied with reduced symmetrical three-phase voltage, while measuring voltage U_{UV}, U_{Vw}, U_{UW}, U_{Vv}, U_{Wv}, U_{Vw}, U_{uv}, U_{uw}, and U_{Vw} at terminals U, u, V, v, W, and v. From the voltages measured in this way, we determine according to Table 9. 1 clock angles for individual terminal connection cases.

9.3.3 Measurement of transmission, voltage, and no-load losses

The voltage conversion of an *unloaded single-phase transformer* is determined by the ratio of the nominal primary voltage U_1 to the voltage on its secondary winding U_{20}

$$p_u = \frac{U_1}{U_{20}}. \tag{9.13}$$

Similarly, the voltage conversions of the individual sections of the *three-phase transformer* are defined, while in some cases, the mean value of these conversions is given.

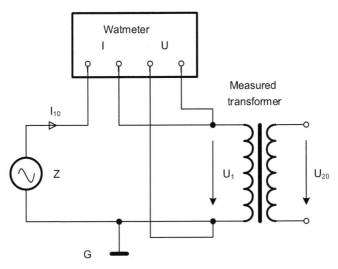

Figure 9.5 Measurement of transmission, no-load of a single-phase phase transformer.

The individual voltage transmissions should not differ by more than 1% with respect to their nominal values. Most transformers have taps on their primary or secondary windings, allowing the transformer gear ratios to change by a few percentages.

The *no-load transformer losses* P_0 are determined by the losses on the resistance of the primary winding R_1 through which the primary no-load current I_{10} flows and the power dissipation in its ferromagnetic core, expressed by the resistance R_{Fe}

$$P_0 = R_1 \cdot I_{10}^2 + R_{Fe} \cdot I_{Fe}^2 \, [W]. \tag{9.14}$$

The resistance losses of the transformer primary winding are significantly smaller in power HV transformers than the losses in their ferromagnetic cores.

Figure 9.5 shows the circuit diagram for the conversion, no-load of a single-phase transformer.

When measuring the no-load transformer properties, its primary voltage first gradually increases up to 1.2 times its nominal value. Then this voltage decreases by 10% of the nominal value to its nominal value. From the measured no-load current I_{10} and voltage U_1, the active power in the primary circuit of the transformer

$$P_{10} = U_1 \cdot I_{10} \cdot \cos\varphi_{10}, \tag{9.15}$$

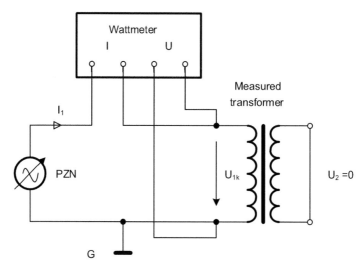

Figure 9.6   Measurement of short-circuit transformer voltage and losses.

where φ_{10} is the phase angle between the voltage phasor U_1 and the current I_{10}. This power corresponds to the power dissipation P_0 in the primary winding of the transformer and the losses in its ferromagnetic core.

9.3.4 Measurement of short-circuit voltages and losses

The *short-circuit voltage* of the transformer, i.e., at its short-circuited output, is defined by the voltage at its primary winding U_{1k} at which the rated current I_1 flows through this winding.

The *short-circuit transformer losses* are determined by the loss on resistance R_1 of the transformer primary winding through which the rated primary current I_1 flows, the loss on resistance R_2 through which the secondary short-circuit current I_{2k} flows, and the losses in its ferromagnetic core, expressed by R_{Fe}

$$P_k = R_1 \cdot I_1^2 + R_{Fe} \cdot I_{Fe}^2 + R_2 \cdot I_{2k}^2 \,[\mathrm{W}]. \tag{9.16}$$

Figure 9.6 is a circuit diagram for measuring short-circuit transformer voltages and losses.

When measuring the voltage of a transformer for a short time, its primary voltage gradually increases to a value at which its primary current reaches its nominal value, while the measurement must take a minimum of time, because the transformer overheats disproportionately during longer measurements and may be damaged. The active power in the primary and secondary windings of the transformer, including the losses in its ferromagnetic core, is determined from the measured current I_1 and the short-circuit voltage U_{k1}.

$$P_{1k} = U_{1k} \cdot I_1 \cdot \cos\varphi_{1k}. \tag{9.17}$$

where φ_{1k} is the phase angle between the voltage phasor U_{1k} and the current I_1. This power corresponds to the power dissipation P_k in the primary and secondary windings of the transformer and the losses in its ferromagnetic core.

9.3.5 Measuring the efficiency of transformers

The basic connection for measuring the efficiency of single-phase transformers is shown in Figure 9.7.

The *efficiency of the measured transformer*, referred to as cos φ, is determined by the ratio of its active output power P_2 to the active input power P_1

$$\cos\varphi = \frac{P_2}{P_1} = \frac{U_2 \cdot I_2 \cdot \cos\varphi_2}{U_1 \cdot I_1 \cdot \cos\varphi_1}, \tag{9.18}$$

where φ_1 and φ_2 are the phase shifts of the phasors of its primary and secondary voltages U_1 and U_2 with respect to its primary and secondary currents I_1 and I_2. The efficiency of large transformers reaches up to 98%, while their efficiency increases with their rated power.

Previously, separate ferromagnetic *voltmeters, ammeters*, and *wattmeters* were used to measure transformer voltages, currents, and losses, which are now being replaced by electronic digital wattmeters, in which voltage and current voltage and current waveforms are usually digitized by 16-bit ADCs with several sampling frequencies of the kSps. From the samples obtained in this way, not only the effective values of voltages and currents are determined but also the active, reactive, and apparent powers, including the power factor (Figure 9.8).

Assuming that the voltage and current waveforms are sampled at time-identical intervals with the period T_s, then the RMS values of voltage and current are expressed by the following relations (Figure 9.9):

$$U = \sqrt{\frac{1}{N}\sum_{i=1}^{N} \cdot N_u^2(t_i)}, \qquad I = \sqrt{\frac{1}{N}\sum_{i=1}^{N} \cdot N_i^2(t_i)}. \tag{9.19a, b}$$

The *active power* is determined by the mean value of the product of the numerical equivalents of the voltage $N_u(t_i)$ and the currents $N_u(t_i)$ at the same time moments t_i

$$P = \frac{1}{N}\sum_{i=1}^{N} N_u(t_i) \cdot N_i(t_i). \tag{9.20}$$

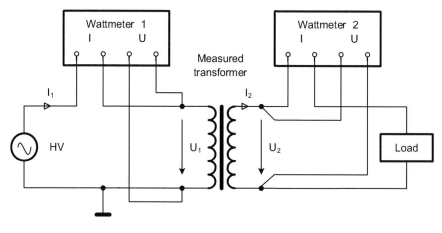

a) wiring diagram

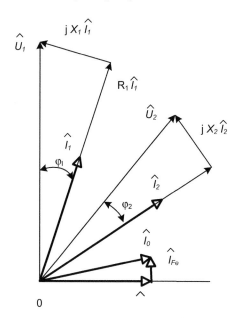

b) phasor diagram of measured quantities

Figure 9.7　Measuring the efficiency of a single-phase transformer.

Reactive power is determined by the mean value of the product of numerical equivalents of voltage $N_u(t_i)$ and currents $N_i(t_i - T_p/4)$, time-shifted by a quarter of the period T_p with respect to voltage samples

$$Q = \frac{1}{N} \sum_{i=1}^{N} N_u(t_i) \cdot N_i \left(t_i - \frac{T_p}{4} \right). \tag{9.21}$$

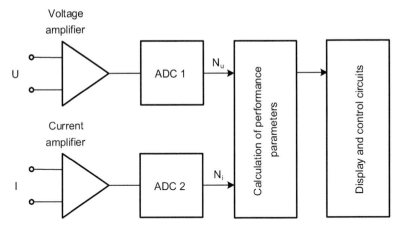

Figure 9.8 Block diagram of a single-phase digital wattmeter.

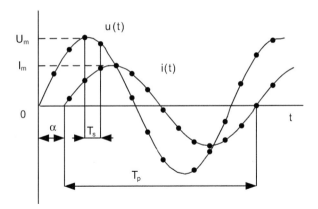

Figure 9.9 Sampled voltage and current waveforms.

Apparent power is determined by the product of the effective values of voltage and current

$$S = \sqrt{\frac{1}{N}\sum_{i=1}^{N}N_u^2(t_i)} \cdot \sqrt{\frac{1}{N}\sum_{i=1}^{N}N_i^2(t_i)} = \frac{1}{N}\sqrt{\sum_{i=1}^{N}N_u^2(t_i) \cdot \sum_{i=1}^{N}N_i^2(t_i)}. \quad (9.22)$$

The *efficiency of a single-phase transformer* is determined from the ratio of active and apparent power

$$\eta = \frac{P}{S}. \quad (9.23)$$

The accuracy of the calculation of individual quantities depends on the number of N samples and the nature of the voltage and current. At sampling

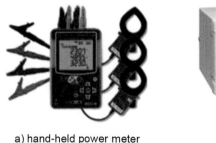 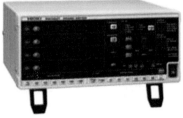

a) hand-held power meter   b) laboratory power meter

Figure 9.10   Digital power meters.

frequencies of several tens to hundreds of ksps, the accuracy of determining these quantities can be achieved up to several hundreds of percentage.

Figure 9.10 shows a typical *three-phase digital wattmeter* KEV 6305 from KYORITSU, allowing to measure voltage up to 600 V, current up to 600 A, active, reactive, and apparent power including power factor with an accuracy of several percentages, and *laboratory three-phase power meter* PW3337 from HIOKI, enabling to measure voltage up to 600 V and current up to 65 A with an accuracy up to 0.1%.

Instrument measuring voltage and current transformers are used to extend the voltage and current ranges of power meters. *Instrument voltage measuring transformers* are manufactured with nominal input voltages of 6–110 kV with an accuracy of 0.1%–0.5% with a nominal output voltage of 100 V. *Instrument current measuring transformers* are manufactured with rated input currents of 5 A to 50 kA with an accuracy of 0.1%–1% and a rated output current of 5 or 1 A.

Figure 9.11 shows a voltage measuring transformer up to 110 kV and a set of CT687× bushing current transformers from HIOKI, enabling to measure currents up to 2 kA.

Figure 9.12 is a circuit diagram for measuring the power quantities of a single-phase transformer with a voltage and current measuring transformer in its primary circuit.

If the conversion ratios of voltage and current transformer are p_u and p_i, then for the measured current and voltage in the primary circuit of the measured transformer,

$$U_1 = \frac{U_m}{p_u}, \qquad I_1 = \frac{I_m}{p_i}. \qquad\qquad (9.24a, b)$$

It applies to active, reactive, and apparent power in the primary circuit of the measured transformer

a) measuring voltage transformer up to 110 kV   b) set of flow current measuring transformers

Figure 9.11 Voltage and current measuring transformers.

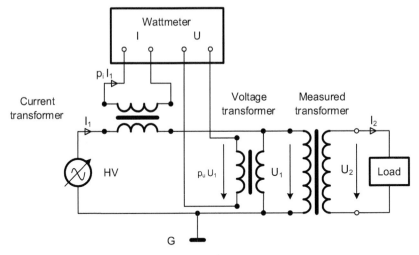

Figure 9.12 Measurement of power quantities of a single-phase transformer with a measuring transformer.

$$P_1 = U_1 \cdot I_1 \cdot \cos\varphi_1 = \frac{U_m}{p_u} \cdot \frac{I_m}{p_i} \cos\varphi_1, \quad Q_1 = \frac{U_m}{p_u} \cdot \frac{I_m}{p_i} \cdot \sin\varphi_1, \quad (9.25\text{a, b})$$

$$S_1 = U_1 \cdot I_1 = \frac{U_m}{p_u} \cdot \frac{I_m}{p_i} \quad (9.25\text{c})$$

where φ_1 is the angle between the primary voltage and current phasors.

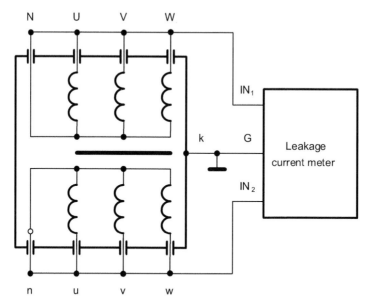

Figure 9.13   Connection for measuring the insulation resistance between the primary and secondary windings of a three-phase transformer.

9.3.6 Measurement of insulation resistance, polarization index, and transformer time constant

Measuring the insulation resistance between the individual windings and between the windings and the transformer frame makes it possible to detect the weakest points of its insulation system, which are usually caused by the presence of moisture and conductive particles in this system.

HV meters are used to measure the insulation resistance of transformers and the actual measurement is performed with a DC voltage of 2.5 kV under normal conditions, i.e., at a relative humidity of 90% and a temperature of 20°C.

Figure 9.13 shows the connection for measuring the insulation resistance between the primary and secondary windings of a three-phase transformer.

In order to remove the residual charge in the transformer insulation system, all its windings must be short-circuited for at least 5 minutes and connected by a grounded earth, or transformer vessel. The windings whose influence is to be excluded shall be earthed. When measuring the insulation resistance repeatedly, the same phase sequence, connection method, and polarity of the measuring voltage must be observed.

Table 9.2 Connection of transformer windings when measuring their insulation resistances (HV – high voltage, MV – medium voltage, LV – low voltage).

Transformer with double winding		Triple winding transformer	
Measured winding	**Grounded winding**	**Measured winding**	**Grounded winding**
HN	LV	HV	SN, NN
LV	HV	MV	HV, MV
HV + LV	Ground	LV	HV, MV
		HV + MV	LV
		HV + MV + LV	Ground

The *polarization index* of the insulation is determined by the ratio of the insulation currents I_{15} and I_{60}, read after 15 and 60 s after connecting the measuring voltage

$$p_i = \frac{I_{15}}{I_{60}}. \qquad (9.26)$$

The *time constant* of the transformer insulation system is determined by the product of its insulation resistance $R_{iz}(60)$, measured after 60 s after connecting the measuring voltage with the system capacity of $C_{iz}(50)$, determined at the measuring AC voltage frequency of 50 Hz

$$\tau = R_{iz}(60) \cdot C_{iz}(50). \qquad (9.27)$$

The measured values of insulation resistance, polarization index, and time constant are assessed in terms of their difference with previously determined values, e.g., one or more years of transformer operation, or with values measured after its production, or repair. If the differences in values differ by more than 40%, then it is necessary to look for the cause of these differences.

For example, the minimum value of the polarization index for newly manufactured transformers should not be greater than 1.7, and for operated transformers, 1.3. The time constants of fault-free transformers range from 5 to 12 s according to their rated power and test voltage.

Table 9. 2 shows the connection of individual sections of the winding of a three-phase transformer when measuring its insulation resistances.

9.3.7 Measurement of loss factor and winding capacity

The *loss factor* expresses the active losses in the isolation of transformers, which are caused by the polarization of its dielectric. As the loss factor increases, the insulation heats up, which leads to its aging and thus to a reduction in its insulating strength. The moisture content of the insulation

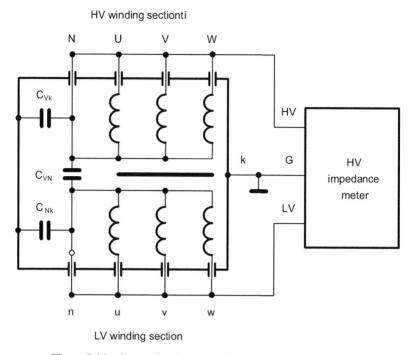

Figure 9.14 Connection for measuring transformer capacitances.

has a major influence on the size of the loss factor or occurrence of partial discharges in isolation.

The loss factor of transformers is measured at test voltages given by 1.2 times their nominal voltages in the temperature range of 20‾30°C.

Figure 9.14 shows a circuit for measuring the capacitances and loss factors of a transformer with one HV and one LV winding, where the C_{VN} capacity represents the capacitance between these winding sections and the C_{Vk} and C_{Nk} capacitances represent the capacitances between these windings relative to the transformer ground.

The measurement consists of three steps. In the *first measurement step*, the low-voltage LV winding is grounded and the measured capacitance C_{m1} is thus determined by the parallel connection of the capacitances C_{Vk} and C_{VN}

$$C_{m1} = C_{Vk} + C_{VN}. \tag{9.28}$$

In the *second measurement step*, the high-voltage HV winding is grounded and the measured capacity C_{m2} is thus determined by the parallel connection of the capacities C_{Nk} and C_{VN}

$$C_{m2} = C_{Nk} + C_{VN}. \tag{9.29}$$

In the *third measurement step*, both windings are interconnected and the measured capacity C_{m3} is thus determined by the parallel connection of capacities C_{Nk} and C_{Vk}

$$C_{m3} = C_{Nk} + C_{Vk} \tag{9.30}$$

From these equations, the winding capacitances to ground and to each other can be determined

$$C_{Vk} = \frac{C_{m1} - C_{m2} + C_{m3}}{2}, \quad C_{Nk} = \frac{C_{m2} - C_{m1} + C_{m3}}{2}, \quad C_{VN} = \frac{C_{m1} + C_{m2} - C_{m3}}{2},$$

$$\text{(9.31a, b, c)}$$

and their loss factors

$$tg\delta(C_{Vk}) = \frac{C_{Vk} \cdot tg\delta(C_{Vk}) + C_{VN} \cdot tg\delta(C_{VN}) - C_n \cdot tg\delta(C_{Nk})}{2C_{Vk}}, \tag{9.32a}$$

$$tg\delta(C_{VN}) = \frac{C_{VN} \cdot tg\delta(C_{VN}) + C_{Nk} \cdot tg\delta(C_{Nk}) - C_{Vk} \cdot tg\delta(C_{Vk})}{2C_{VN}}, \tag{9.32b}$$

$$tg\delta(C_{Nk}) = \frac{C_{Vk} \cdot tg\delta(C_{Vk}) - C_k \cdot tg\delta(C_{Nk}) + C_{VN} \cdot tg\delta(C_{VN})}{2C_{Nk}}, \tag{9.32c}$$

A similar methodology is used to measure not only the capacitances and loss factors of the insulation system of *multi-winding transformers* but also the insulation systems of stators and rotors of rotating electrical machines with multiple windings.

Because these measurements are highly sensitive to external interfering electromagnetic fields, it is recommended to perform the measurements even when replacing the measuring cables to the transformer winding and the resulting capacity and determine their losses using the following equations:

$$C_s = \frac{C_{x1} + C_{x2}}{2}, \qquad tg\delta_s = \frac{C_{x1} \cdot tg\delta_{x1} + C_{x2} \cdot tg\delta_{x2}}{C_{x1} + C_{x2}}. \tag{9.33a, b}$$

Due to the fact that the loss factors of transformer insulation are temperature-dependent, a temperature coefficient related to a temperature of 20°C is used to compare them.

$$k_\vartheta = \frac{tg\delta_\vartheta}{tg\delta_{20}}, \tag{9.34}$$

where $tg\delta_\theta$ is the insulation loss factor at temperature θ.

Table 9.3   Temperature coefficients of insulation loss factor impregnated with transformer oil.

ϑ (°C)	10	15	20	25	30	35	40	45	50	55	60	65	70
	0.80	0.90	1.00	1.12	1.25	1.40	1.55	1.75	1.95	2.18	2.42	2.7	3.0

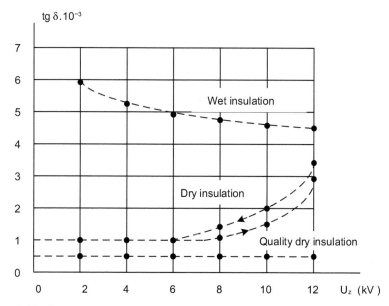

Figure 9.15   Dependence of the insulation coefficient of the HV transformer 100 kVA on the test voltage.

Table 9. 3 shows the temperature coefficients of the loss factor of paper insulations impregnated with transformer oil.

Figure 9.15 shows the typical dependences of the HV transformer insulation loss factor on the test voltage for its high-quality non-wet and high-quality dry and wet insulations.

It is clear from the above dependences that the loss factor of quality *dry insulation* is very small and practically independent of the test voltage. In *poor quality dry insulation*, the losses in the dielectric increase from a certain voltage due to the ionization of the dielectric and the increase in partial discharges in it. These phenomena cause hysteresis of voltage dependence losses. The loss factor of the *wet insulation gradually* decreases not only with increasing test voltage but also with the measurement time, as the active dielectric losses in the insulation gradually dry it out. Therefore, in order to measure the loss factor correctly, the transformer should be placed in a dry environment, or its insulation is to be dried, e.g., by its operation in a loaded condition.

Table 9.4 Maximum values of transformer loss factor in percentage at 50 Hz.

Transformer status	Operating temperature [°C]						
	10	20	30	40	50	60	70
Before commissioning							
to 35 kV and power < 10 MVA	1.2	1.5	2.0	2.6	3.4	4.5	6.0
to 35 kV and power > 10 MVA	0.8	1.0	1.3	1.7	2.3	3.0	4.0
After repair or revision							
to 35 kV	2.5	3.5	5.5	8.0	11.0	15.0	20.0
to 110 kV (150 kV)	2.0	2.5	4.0	6.0	8.0	12.0	16.0

Table 9.5 Limit values of the C_2/C_{50} ratio.

Temperature [°C]	10	20	30	40	50
C_2/C_{50}		1.1	1.2	1.3	1.4	1.5

In Table 9. 4, the maximum permissible loss factors of transformers in terms of their power and operating temperatures are given.

To assess the condition of HV insulation of electrical machines, it is important to respect not only the effect of temperature but also data on its previous condition. The capacity of the insulation decreases with increasing temperature and frequency. Therefore, when diagnosing the condition of the insulation, the ratio of its capacity C_2, measured at 2 Hz, to the capacity C_{50}, measured at 50 Hz, is determined. Typical limit values for this ratio as a function of temperature are given in Table 9. 5.

9.3.8 Winding fault detection

Frequency response analysis (FRA) frequency impedance response methods are used to detect transformer winding faults, which analyze changes in this impedance caused by winding faults, such as winding position changes and interthread shorts.

Either the *impulse frequency response analyzing method (IFRA)* or the *sweep frequency response analyzing method (SFRA)* is used to determine the transformer transmission impedance.

The result of both methods is to determine the frequency dependence of the transmission impedance of the tested transformer, from which the extent of its damage can be determined after comparison with the frequency dependence of the transmission impedance of a transformer with an intact winding.

Figure 9.16 shows a replacement diagram of the transformer with part of its winding, ferromagnetic core, and oil bath.

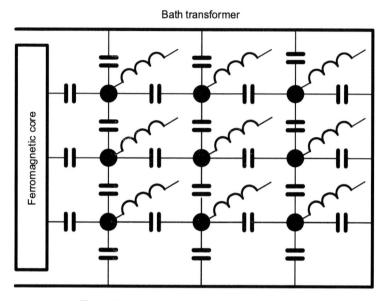

Figure 9.16   Transformer replacement diagram.

It is clear from the figure that the transformer replacement diagram represents a circuit with spread parameters that represent the resistances of the conductors, their parasitic capacitance, and inductance not only between them but also between them and the ferromagnetic core of the transformer and its bath.

The *transmission impedance* of the transformer is given by the ratio of the difference of the voltage phasors between its primary and secondary windings to the current phasor of the flowing current between these windings.

$$\hat{Z}_p = \frac{\hat{U}_2 - \hat{U}_1}{\hat{I}_2}. \tag{9.35}$$

The *IFRA (impulse frequency response analyzing) method* uses voltage pulse signals $u_1(t)$, which are connected to one of the poles of the primary winding of the transformer, on the secondary winding of which current pulses $i_2(t)$ are sensed and of which the sensing resistor R_s consists of voltage pulses with time course $u_2(t)$ (Figure 9.17).

The *transmission impedance of the transformer* is then given by the following equation:

$$\hat{Z}_p = \frac{\hat{U}_2 - \hat{U}_1}{\hat{U}_2} R_s, \tag{9.36}$$

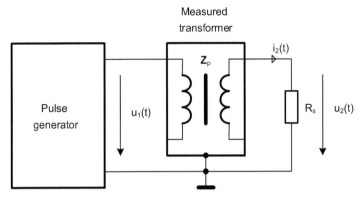

Figure 9.17 Transformer transmission impedance measurement by pulse method.

where $\widehat{U}_1$ and $\widehat{U}_2$ are phasors of frequency spectra of voltage waveforms $u_1(t)$ and $u_2(t)$ obtained by *Fourier transform*

$$\widehat{U}_1(f) = \frac{1}{T}\int_{-\infty}^{\infty} u_1(t)\exp(-j\cdot 2\cdot\pi\cdot f\cdot t), \quad \widehat{U}_2(f) = \frac{1}{T}\int_{-\infty}^{\infty} u_2(t)\exp(-j\cdot 2\cdot\pi\cdot f\cdot t).$$

$$(9.37a, b)$$

The transmission impedance of a transformer is usually expressed by its *amplitude* and *phase frequency response*, defined by the following equation:

$$\widehat{Z}_p = \left|\widehat{Z}_p\right|.e^{j\varphi},\qquad (9.38)$$

where $\left|\widehat{Z}_p\right|$ is the absolute value of the impedance and φ is the phase shift between its real and imaginary components.

Often, instead of the frequency dependence of the impedance, the frequency dependence of the transformer transmission attenuation is displayed, which is defined by the logarithmic ratio of the absolute values of the voltage phasors at the nominal value of the sensing resistor.

$$B = 20.\log\frac{|U_2|}{|U_1|}.\qquad (9.39)$$

Since the wave impedance of the coaxial supply cables is usually 50 Ω, , the value of the sensing resistance is also selected as 50 Ω in terms of non-reflective current sensing.

Figure 9.18 shows typical transmission attenuation waveforms of a transformer with an intact winding and a transformer with a damaged winding.

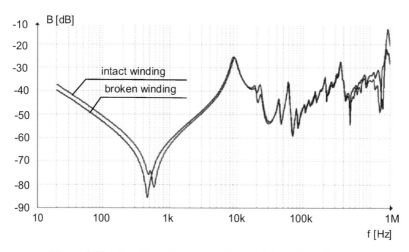

Figure 9.18   Amplitude frequency characteristics of transformer.

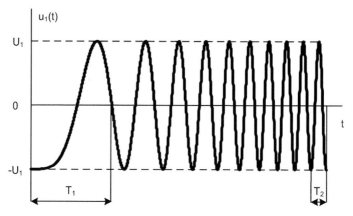

Figure 9.19   Time-shifted sinusoidal voltage.

The *sweep frequency response analyzing (SFRA)* frequency analysis method uses a frequency-amplified constant-amplitude sinusoidal signal (*chirp signal*) to determine the transformer transmission impedance.

$$u_1(t) = U_1 \cdot \sin\left(2 \cdot \pi \cdot f_2 - 2 \cdot \pi \cdot f_1\right)t, \tag{9.40}$$

where $f_1 = 1/T_1$ and $f_2 = 1/T_2$ are the frequencies determining the frequency range of this signal (Figure 9.19).

The result of this analysis is, similarly to the previous method, the amplitude-frequency characteristic of the transmission attenuation of the tested transformer.

The advantage of the SFRA method over the IFRA method is its greater achievable accuracy in determining the frequency dependence of the transformer transmission impedance and a simpler way of processing the results.

The following methods are used to evaluate winding faults due to changes in the frequency characteristics of the transformer transmission impedances.

In the simplest case, a method is used that determines the differences in areas below the frequency characteristics of the transmission impedances B_1 and B_2 of transformers with intact and broken windings. This difference in areas is expressed by a coefficient

$$k_f = \int_{f_1}^{f_2} \left[B_1(f) - B_2(f) \right]^2 df.$$ (9.41)

For $k_f = 1$, the state of the inspected transformer is identical to its previous state; at $k_f < 10$, the transformer operates in normal mode, at $10 < k_f < 20$, another measurement must be performed at the next inspection, and at $k_f < 20$, it is necessary to perform a visual inspection of the winding, possibly its correction.

A frequently used and internationally recommended IEEE standard method for evaluating the difference in frequency characteristics of transformer transmission is to determine the correlation between the two frequency characteristics of attenuation, which is expressed by a *correlation coefficient*

$$r = \frac{\sum_{i=1}^{n} (x_i - x_s) \cdot (y_i - y_s)}{\sqrt{\sum_{i=1}^{n} (x_i - x_s)^2 \cdot \sum_{i=1}^{n} (y_i - y_s)^2}},$$ (9.42)

where

$$x_s = \frac{1}{n} \sum_{i=1}^{n} x_i, \qquad y_s = \frac{1}{n} \sum_{i=1}^{n} y_i$$ (9.43a, b)

are the *mean values* of the frequency and transmission attenuation samples and n is their number of samples.

The advantage of this evaluation method is that it is highly sensitive to changes in local resonance points of the frequency characteristics of transformer transmissions, respectively to their extinction or occurrence. Its disadvantage is that it is not able to detect changes caused by a simple shift of frequency characteristics in a given frequency band.

Furthermore, the *average absolute deviations* are used to express the difference in the frequency characteristics of the attenuation of the transformers

$$r = \frac{1}{n} \sum_{i=1}^{n} |x_i - y_i|,$$

(9.44)

and *mean absolute logarithmic deviations*

$$r = \frac{1}{n} \sum_{i=1}^{n} |20\log x_i - 20\log y_i|.$$

(9.45)

The *average absolute logarithmic deviation* very well identifies changes in frequency characteristics with a constant difference over the entire frequency band. Its disadvantage is that it is less sensitive to amplitude changes of the evaluated waveforms.

A coefficient is used to express the differences between the minima and maxima of the frequency characteristics, which are caused by local resonances in the transformer windings

$$MM = \frac{\sum_{i=1}^{n} \min(x_i, y_i)}{\sum_{i=1}^{n} \max(x_i, y_i)}.$$

(9.46)

Assessing changes in the transformer winding by FRA methods cannot be realized on the basis of one measurement, but it is necessary to perform several time-lapse comparative measurements of its transmission impedances and then estimate the degree of damage from the difference of measured data.

A typical instrument for measuring the frequency characteristics of transformer transmission impedances is, for example, an FRA analyzer from OMICRON, which enables SFRA diagnostics in the frequency range of 10 Hz to 20 MHz with a frequency-scattered 1-V signal at a load of 1 MΩ or 50 Ω. The frequency response is measured at a maximum of 3.201 points. The dynamic range of transmission attenuation measurement is greater than 120 dB (Figure 9.20a). A similar device was developed at the Department of Measurement, Faculty of Electrical Engineering, Czech Technical University in Prague (Figure 9.20b).

9.4 Diagnostics of Transformer Oil Properties

The quality assessment of transformer oils is carried out in accordance with the IEC 60 422 standard, which gives detailed instructions for the inspection and maintenance of transformer oils. When assessing the quality of transformer oils, their electrical, mechanical, and chemical properties are determined.

a) OMICRON FRA device   b) CTU FEE FRA device

Figure 9.20   FRA measuring instruments.

9.4.1 Measurement of electrical properties of transformer oils

The electrical properties of transformer oils include their electrical strength, internal resistivity, permittivity, and loss factor.

9.4.1.1 Measurement of breakdown voltage at mains frequency

Determination of electrical strength and breakdown voltage of transformer oils is carried out in accordance with standard EN 60156. This test uses a test chamber with a volume of 350–600 ml, in which hemispherical polished steel electrodes with a diameter of 36 mm at a distance of 2.5 ± 0.05 mm (Figure 9.21).

When filling the chamber with oil, it is necessary to ensure the absence of air bubbles, which would significantly reduce the breakdown voltage of the measured oils.

The electrical strength test is performed in successive six breakthroughs at 5-minute intervals, determining the mean value of the breakdown voltages and their standard deviation.

$$U_{ps} = \frac{1}{6}\sum_{i=1}^{6} U_{pi}, \qquad \sigma = \sqrt{\frac{1}{6}\sum_{i=1}^{6}\left(U_{pi} - U_{ps}\right)^2}. \qquad (9.47a, b)$$

If the standard deviation is greater than 20%, the measurement must be repeated. If, even after repeated measurements, the standard deviation does not fall below 20%, then the oil as an insulator does not comply.

A typical instrument for measuring the electrical strength of not only solid but also liquid insulators is, for example, the Baur DTA 100C

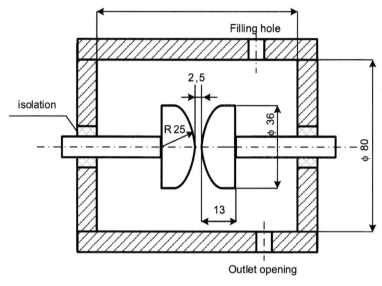

Figure 9.21   Test chamber for measuring the breakdown voltage of liquid insulators.

Figure 9.22   Baur DTA 100C breakdown
voltage meter.

instrument, which allows to measure breakdown voltages up to 100 kV
with an adjustable time change in the range of 500 V/s to 10 kV/s
(Figure 9.22).

9.4.1.2 Measurement of relative permittivity and loss factor

The determination of the relative permittivity and loss factor of transformer
oils is carried out in accordance with the standard EN 60 247. In this test, a

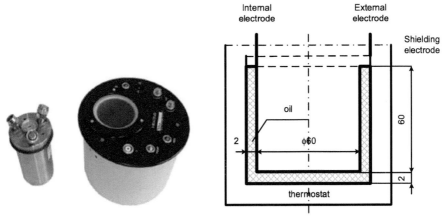

a) test chamber with thermostat   b) section through the chamber with thermostat

Figure 9.23   Test chamber for measuring permittivity and oil loss factor (Tettex 2903).

special test heated cylindrical chamber with inner and outer cylindrical electrodes is used (Figure 9.23).

First, the capacitance C_0 between the two unencapsulated electrodes,
which is approximately 60 pF, is measured. Then a liquid with known permittivity ε_n is poured between the electrodes, and the capacitance C_n between
the electrodes is measured. The so-called *electrode constant* of the chamber
is determined from these data

$$C_e = \frac{C_n - C_0}{\varepsilon_n}$$ (9.48)

and its correction capacity

$$C_g = C_0 - C_e.$$ (9.49)

Then the measured oil is poured between the electrodes and the capacitance
C_x and the loss factor $tg\delta_x$ are measured again. The relative permittivity of the
oil is determined from these values

$$\varepsilon_x = \frac{C_x - C_g}{C_e}.$$ (9.50)

If the oil fully fills the interelectrode system of the chamber, then the correction capacity $C_g = 0$, and for the relative permittivity of the oil applies

$$\varepsilon_x = \frac{C_x}{C_0}.$$ (9.51)

Since the relative permittivity of air and its loss factor is approximately equal to 1, it applies to the loss factor of the measured oil

$$tg\delta_x = \sqrt{\varepsilon_x \frac{C_0}{C_x} - 1}. \tag{9.52}$$

The measurement of both quantities is performed in the frequency range of 40–62 Hz with electrical stress of the oil in the range of 30 V/mm to 1 kV/mm.

9.4.1.3 Measurement of internal resistivity of transformer oils

The measurement of oil resistivity is carried out in accordance with the standard EN 60247. The same test chambers are used in this test as in the measurement of the loss factor and permittivity. The resistivity measurement is performed with a direct voltage so that the voltage stress of the oils is 250 V/mm for 60 ± 2 s. If the temperature dependence of the oil resistivity is to be measured, it must always wait at least 10 minutes to stabilize the temperature. During repeated measurements, the measured values of resistivities should not differ by more than 35%.

The resistivity of oils is defined by the following equation:

$$\rho = k \frac{U}{I} [\Omega \cdot m], \tag{9.53}$$

where k is the constant of the electrode system of the test chamber. This is determined from the resistance measured between the electrodes of the test chamber in which there is a solution of known resistivity, e.g., distilled water, which has a resistivity $\rho_n = 2.2710^{11}$ [$\Omega \cdot m$]. Knowing this resistivity, it applies to the correction factor

$$k = \rho_n \frac{I_n}{U_n}, \tag{9.54}$$

where U_n and I_n are the voltages and currents at the electrodes of the test chamber.

The resistivity of the oil is then determined by the following equation:

$$\rho = \frac{\rho_n}{R_n} \cdot \frac{U}{I}. \tag{9.55}$$

Special HV meters are available to measure the electrical properties of oils. A typical representative of these devices is, for example, the oil and particulate analyzer Tettex 2830 (Figure 9.24).

With the Tettex 2903 and Tettex 2914 test chambers, this device allows to measure resistances in the range of 120 kΩ to 5 TΩ and the resistivity of

Figure 9.24   Tettex 2830 oil and solid dielectric analyzer.

solid and liquid insulators in the range of 900 kWm to 33 TWm at DC voltages of 250–2500 V. It further allows to measure capacitance in the range of 10 pF to 1 mF, permittivity of solid and liquid insulators in the range of 1–30, and their loss factor in the range of 0–100 at AC voltage of 40–2500 V in the frequency range of 40–65 Hz.

9.4.2 Measurement of physical properties of transformer oils

The monitored physical properties of transformer oils include their color, density, viscosity, flash points, burning, and solidification.

9.4.2.1 Transformer oil color measurement

The color of transformer oils depends on the degree of their oxidation, which is caused by the polymerization and polycondensation of oxygen products in the oil. The color changes of the oils thus determine the degree of their oxidative aging, which in turn reduces their electrical strength and increases the loss factors. Colorimeters are used to determine the color shade of transformer oils, in which the color of the test oil is compared with the color scale. Modern colorimeters use color *CCD sensors* to compare colors.

Figure 9.25 shows the LICO 860 spectral colorimeter from HACH, enabling the measurement of liquid colors in the light wavelength range of 380–740 nm with a resolution of 10 nm.

Figure 9.26 shows a demonstrative range of colors according to the IDSO 2049 standard.

Figure 9.27 shows the color changes of the transformer oil at the time of its use.

9.4.2.2 Transformer oil density measurement

Although the density of transformer oils does not belong to an important diagnostic quantity, it is used for the basic classification of transformer oils.

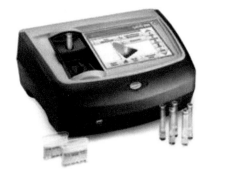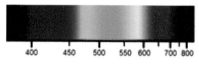

Figure 9.25   LICO 860 spectral colorimeter from HACH.

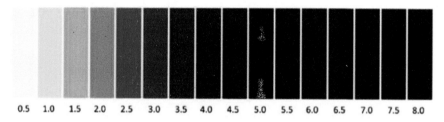

Figure 9.26   Color range according to ISO 2049.

Figure 9.27   Example of oil color change during its operation.

The density of transformer oils is usually given by the weight of 1 ml of oil in grams.

Either float or refractometric density meters are used to measure density.

The *float densitometer* consists of a glass tube in which a float moves, which is made of a material whose specific gravity is always less than the minimum density of the oil being measured. The density of the oil is then determined by the size of the surfaced part of the float above its surface.

The *refractometric densitometer* uses different refractive indices of the measured oil and air. When the light beam strikes the oil surface, the beam is partially reflected at an angle whose value is directly proportional to the

a) float densitometer   b) refractometric densitometer

Figure 9.28   Transformer oil densitometers.

density of the oil being measured. The angle of reflection is read in the optical system of the refractometer.

Figure 9.28 shows typical densitometers used to measure the density of transformer oils.

9.4.2.3 Measuring the viscosity of transformer oils

Viscosity of transformer oils, indicates the ratio between the tangential voltage τ and the change in the flow rate v/dt of adjacent layers of liquids

$$\eta = \frac{\tau}{\frac{dv}{dt}} \left[N \cdot s \cdot m^{-2} \right]. \tag{9.56}$$

Because the viscosity of oils increases significantly with their temperature, its determination is important in terms of cooling efficiency of transformers.

Rotary, capillary, and body viscometers are used to measure the viscosity of oils.

The *rotary viscometer* determines the viscosity of liquids by the moment of force M, which must overcome the rotating body immersed in the liquid

$$M = k \cdot \omega \cdot \eta, \tag{9.57}$$

where n is the number of revolutions of the rotating body, and k is a constant depending on its geometric shape. The number of revolutions of the rotating body must be such that there is no turbulent flow between this body and the container filled with oil.

The *capillary viscometer* works on the principle of measuring the volume flow of the measured liquid through a tube of defined dimensions. If the tube has a radius r and a length l, then a liquid of volume V flows through it. The *Hagen–Poiseuill equation* applies to the viscosity of the liquid

$$\eta = \frac{\pi \cdot r^3 \cdot h \cdot \rho \cdot g \cdot t}{8 \cdot V \cdot l} \left[N \cdot s \cdot m^{-2} \right], \tag{9.58}$$

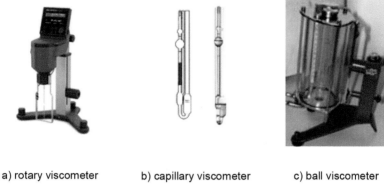

a) rotary viscometer   b) capillary viscometer   c) ball viscometer

Figure 9.29   Transformer oil viscometers.

where h is the height above the ground from which the liquid flows, ρ is its density, and $g = 9.81$ [ms^{-2}] is the acceleration of the earth.

Body viscometers are based on measuring the immersion speed or, conversely, the emergence of a test float to the liquid level. The force acting on a flowing float is determined by *Stokes' law*, which has a magnitude for a commonly used spherical float of radius r

$$F = 6 \cdot \pi \cdot \eta \cdot r \cdot v \, [\mathrm{N}]. \qquad (9.59)$$

Figure 9.29 shows a typical rotary, capillary, and float viscometer.

9.4.2.4 Measurement of flash and combustion temperatures

These parameters are determined mainly from the point of view of fire safety.

The oil *flash point* is defined by the temperature at which the test flame approaches the oil container and then goes out. The flash point of transformer oils should be higher than 150°C. The *burning point* of the oil is the temperature at which the flame of the already ignited oil no longer goes out.

A *Pensky–Martens instrument* with a gas test flame is used to determine both points. Each flash of oil is manifested by a small explosion above the oil level and an accompanying flash, during which the oil temperature is measured. When measuring these quantities, the temperature, pressure, and humidity of the surrounding air are also recorded.

9.4.2.5 Freezing temperature measurement

This parameter is especially important when commissioning transformers at low temperatures and low loads. The solidification temperature is determined by the temperature at which the oil begins to solidify and thus does not perform either insulating or cooling tasks. The solidification temperature of oils

Table 9.6 Properties of transformer oils.

Parameter	Unit	Mineral oil OMV trafo IEC II	Vegetable oil Envitotemp FR3	Synthetic oil Midel 7131
Resistivity	$\Omega \cdot m$	$2 \cdot 10^{11}$	10^{10}	$5 \cdot 10^{12}$
Permittivity	$F \cdot m^{-1}$	2.1	1.8	2.0
Loss factor		10^{-4}	$2 10^{-4}$	$3 10^{-5}$
Electrical strength	$kV \cdot mm^{-1}$	30	20	35
Relative density	$kg \cdot dm^{-3}$	0.862	<0.96	0.97
Point of ignition	°C	145	310–320	275
Viscosity at 40°C	$mm^2 \cdot s^{-1}$	9.5	< 50	28
Viscosity at –20°C	$mm^2 \cdot s^{-1}$	850	–	1500
Freezing point	°C	<–60	–18 to –24	–60
Acid number	$Mg \ KOH.g^{-1}$	<0.01	<0.06	<0.01
Water content	$mg \cdot kg^{-1}$	10	150	50
PCB contents	$mg \cdot kg^{-1}$	0	0	0

is determined by their gradual cooling until their fluidity ceases, at which point the viscosity of the oil ceases.

Table 9.6 shows the properties of oils used in HV and VHV transformers.

9.4.3 Measurement of chemical properties of transformer oils

The monitored chemical properties of transformer oils include the water content in the oil, the acidity of the oil, the presence of sulfur, and the content of polycyclic aromatics and polychlorinated biphenyls, including gassing of oils.

9.4.3.1 Water content measurement

Water absorption in transformer oils significantly impairs their electrical strength and dielectric losses. Water enters oils mainly from the air or by oxidative aging in polycondensation reactions. The presence of water in oils manifests itself in dissolved, emulsified, and free form.

Dissolved water is created by the absorption of water in the oil and poses a significant risk in the operation of transformers. *Emulsified water* is formed by microscopic drops of water in oil, which form hydrophilic or hydrophobic emulsions. Hydrophilic emulsions are formed when substances are present in the oil, which dissolve in water and not in the oil. *Hydrophibic emulsions* are formed when foreign components in the oil dissolve in it and are insoluble in water. Free water usually occurs at the bottom of transformer tanks.

Figure 9.30 shows the dependence of the electrical strength of mineral transformer oil on the amount of water contained in it.

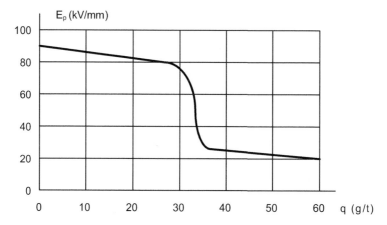

Figure 9.30 Dependence of the electrical strength of mineral oil on the amount of water contained in it.

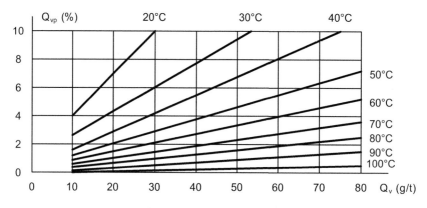

Figure 9.31 Nielsen diagram.

In addition to reducing the electrical strength of the oil itself with increasing water content, there is also the effect of reducing the electrical strength of the solid insulators that receive water from these oils. In the operation of transformers, due to temperature fluctuations, water passes between the solid and liquid insulation systems of the transformers. This phenomenon is described by the so-called Nielsen diagram, which shows the dependence of the amount of water in the cellulose insulation on the water content in the oil and its temperature (Figure 9.31).

The presence of water in transformer oils is assessed in accordance with standard EN 60 076-1 by a *coulometric titration method*, which is based on the reaction of water with iodine with sulfur dioxide, an organic base, and an alcohol dissolved in an organic solvent.

The diagnosed oil sample is first checked for clarity and water and mechanical impurities. The oil sample is then mixed with an alcoholic solution of iodine and sulfur dioxide. The electrolytic reaction of iodine with water creates a charge of 0.72°C from 1 mg of water, which is a measure of the water content of the oil.

The water content can also be determined by a method based on the reaction of a calcium hybrid with water to form hydrogen, the amount of which is proportional to the water content of the oil.

9.4.3.2 Oil acid measurement

According to the EN IEC 62021-1 standard, the acidity of oil is one of the basic diagnostic parameters of oils and thus indicates their aging.

The *acidity of oils* is defined by the amount of potassium hydroxide (KOH) needed to neutralize the acidic reactants in 1 g of oil. According to the acidic substances contained in the liquid, the degree or degree of degradation of the oil is determined. To determine the acidity of the oils, either the potentiometric method or the alkaline blue acidity method 6B is used.

The *potentiometric method* determines the acidity of oils by the amount of potassium hydroxide (KOH), which must neutralize organic and inorganic substances in a non-conductive alcoholic environment to an acid value of PH = 11.5 in 1 gram of oil.

Method for determination of acidity to alkaline blue 6B is based on the titration of acidic compounds contained in the oil using an alcoholic solution of potassium hydroxide. In accordance with the ISO 6618 standard, the so-called 0.1 M KOH factor is first determined by adding an alkaline blue solution to a 1:2 mixture of ethanol and toluene until this mixture turns light blue. An alcoholic solution of potassium hydroxide is then added until the initial red color of this solution. 10 mg of benzoic acid are added to the solution thus neutralized, and after dissolving it, the solution thus obtained is titrated with an alcoholic solution of potassium hydroxide, the acidity factor of which is to be determined.

The following then applies to the acidity of the transformer oil

$$PH = \frac{a \cdot f \cdot N \cdot 56,11}{m},\tag{9.60}$$

where f is the factor of 0.1 M KOH and is the consumption of the KOH sample, m is the weight of the oil in grams, and N is the *molarity of the titrated alcoholic* KOH solution.

The acidity of oils is related to their saponification number, which indicates the amount of potassium hydroxide needed to *saponify* the substances in the oil.

9.4.3.3 Measurement of the presence of sulfur

Determining the presence of sulfur in transformer oils is particularly important because sulfur and its solutions, especially sulfuric acid, cause corrosion of steel and copper. Oxidation methods are used to determine the higher sulfur content of the oil, in which the oil is burned and the sulfur content is determined from the combustion products. X-ray *fluorescence (XRF) spectroscopy* is used for lower sulfur contents, of the order of 10^{-6}.

9.4.3.4 Content of polycyclic aromatics and polychlorinated biphenyls

These substances, contained in some types of transformer oils, have a negative effect on the environment and the human body. These are mainly polycyclic aromatics and polychlorinated biphenyls. *Polycyclic aromatics (PCA)* are detectable by extraction with dimethyl sulfoid (DMSO) by liquid chromatography. *Polychlorinated biphenyls (PCBs)* are detectable by gas chromatography (GC), respectively mass spectrometry.

9.4.3.5 Gas content measurement

The electrical strength of transformer oils, which are mixtures of hydrocarbon molecules of the CH_3, CH_2, and CH groups, is significantly affected by the gases formed during the decomposition of oils and the solid insulation of transformers due to their electrical and thermal stress. During the electrical and thermal stress of the gases, hydrogen H_2, acetylene C_2H_2, and ethylene C_2H_4 are emitted in particular. The main source of gases in transformer oils are partial discharges, in which the conductive channel reaches temperatures of up to several thousand°C.

Although it is rapidly cooled in the oil, it still releases gases, as some of the C-H and C-C hydrocarbon bonds are split. This leads to extensive reactions, which produce unstable hydrocarbon fragments that recombine, especially in hydrogen, ethylene, and acetylene molecules. These gases then dissolve in the oil. For example, for ethylene, these processes begin to occur at temperatures above 500°C, for acetylene between 800 and 1200°C, and among carbon particles at temperatures above 500–800°C.

Figure 9.32 gives an overview of the gas components arising from the electrical and thermal stress of oils.

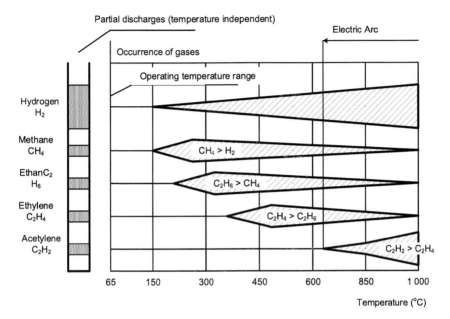

Figure 9.32 Overview of gases released during thermal stress of transformer oils.

The transformers' solid insulators are made up of *polymer chains* containing glucose rings, C-O bonds, and *glycosidic bonds*, which are less thermally stable than the carbon bonds in the oil and therefore decompose at lower temperatures. Their cleavage occurs at temperatures higher than 105°C and complete decomposition and charring occurs at temperatures higher than 300°C. The resulting products of these reactions are water, carbon dioxide (CO_2), carbon monoxide (CO), and minor amounts of other hydrocarbon gases and furan compounds.

The occurrence of carbon dioxide CO_2 and carbon monoxide CO in the cellulose insulation of transformer windings depending on their temperature is shown in Figure 9.33.

References

[1] Hauschild, W., Lemke, E. High Voltage Test and Measuring Techniques. Springer Verlag 2018. ISBN: 978-3-319-97459-0.
[2] Martínek, P.: *Theoretical and experimental analysis of partial discharges in dielectric of electric machines.* Ph.D. work, University in Pilsen, 2005.

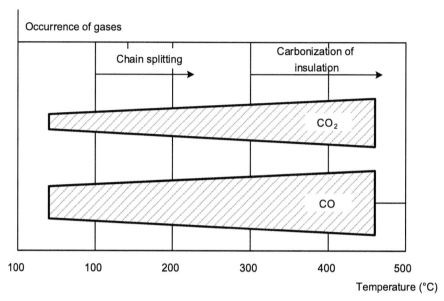

Figure 9.33   Occurrence of CO2 and CO in cellulose insulation of transformer windings.

[3] *Partial dichcharge measurement,* EN 60 270,

[4] Ková Ík, J. *Detection and localization of external sources of sensor discharge in power transformers by acoustic methoda,* Electric engineering, No. 6, 2000,

[5] Schejbal, V. *Transformer diagnostics through electrointelligation fluid analysis,* Ph.D. work, University in Pilsen 2007.

[6] Šomková, M. *Contribution to the diagnosis of power oil transformers.* University of Technology, Faculty of Mechanical Engineering, 2010.

[7] Hindra, M.: *Diagnostic methods of monitoring gases dissolved in transformer oil.* University of Technology, Faculty of Mechanical Engineering, Brno, 2012.

[8] Rozmiler, J. *Design of continuous monitoring system of power transformer,* Bachelor thesis, University in České Budějovice 2010.

[9] Bujaloboková, M., Trnka, P. *Detection of failures of HV machines by analysis of discharge activity with a focus on transformers.* http://dspace.vsb.cz/handle/10084/83878

[10] Mullerová, E. *Non-standard use of ultrasonic localization methods in transformer diagnostics,* Habilitation thesis of University in Pilsen, 2009.

[11] Humlová, V. *Acoustic responses of partial discharges in the diagnosis of insulating defects of power transformers*, Diploma thesis of University in Pilsen, 2011.

[12] Lochman, J. *On-line diagnostics of transformers*, Diploma Thesis of University in Pilsen 2011.

[13] Judd, M. D., Yang, L., Hunter, I. B. B. *Partial Discharge Monitoring for Power Transformers Using UHF Sensors Part 1: Sensors and Signal Interpretation*.

[14] Marakalous, S. M., Wilson, A. *Techniques for detection and location of partial discharges in power transformers*. http://www.doble-lemke.eu/ media/ files/ Media/ Techpapers/ PD/2009/ 2009-01-21

[15] Shen, Z. B., El-Saadany, E. F. *Localization of Partial Discharges Using UHF Sensors in Power Transformers*.

[16] *Guide for the Application and Interpretation of Frequency Response Analysis for Oil-Immersed Transformers*. IEEE-SA Standars Board, Transformers. Committee of the IEEE Power and Energy Society. December 2012

[17] Muhr, M. *Developments in Diagnosis of High Voltage Apparatus*. University of Technology Graz 2009.

10

Diagnostic of Rotary HV Electrical Machines

Rotary electric machines include dynamos, dynamometers, hydro alternators, turbogenerators, synchronous, and asynchronous motors.

10.1 Types of HV Rotary Electrical Machines

Dynamos are rotating electrical machines that generate *direct current*. This current is generated due to electromagnetic induction between the magnetic circuits of the stator and the rotating rotor. A direct current flowing through its winding usually generates the direct current of the stator's magnetic field. Due to the rotation of the rotor in this field, an alternating electric current is induced in the stator winding, which is converted into direct current by means of a *collector brush* located on the dynamo rotor. At present, dynamos are practically not used for the production of electricity, but their principle of operation is used, for example, in *DC motors* and *dynamometers*, of which the other machines are designed to measure the torque and power of rotary machines.

Figure 10.1 shows a dynamo with a commutator and collecting brushes.

In their stator windings due to a rotating magnetic field. The source of this field is either permanent magnets located in the rotor. The most common *three-phase alternators* have stator windings evenly distributed around the stator circuit, and, therefore, a *three-phase alternating current* is induced in them, the components of which have phase shifts of 120°. In terms of alternator rotor design, we distinguish between alternators with a smooth rotor and alternators with excellent poles.

Alternators with non-protruding poles have rotor windings mounted in the rotor grooves, which thus perform the function of poles. Due to the high speed, the rotors of these alternators have a relatively small diameter, approximately up to 1 m, in order to prevent it from being pulled out of the grooves

Figure 10.1   Dynamo with commutator and collecting brushes.

Figure 10.2   Alternator rotor with a smooth rotor with a power of 500 MVA.

due to the centrifugal force acting on the rotating winding. The length of the rotors of these machines reaches up to several meters. These alternators, powered by steam or gas turbines, referred to as turbo alternators, are used in thermal, gas, and nuclear power plants.

Alternators with a smooth rotor and air-cooling are manufactured in the power range from 100 to 350 kVA with a nominal voltage of up to 20 kV. Alternators with combined water-cooling, or nitrogen, is produced up to 1.5 GVA with nominal voltages from 15 to 24 kV.

Figure 10.2 shows the alternator rotor with a smooth rotor having a power of 500 MVA.

Alternators with *protruding rotor poles* have pole pieces mounted around the rotor circumference. The number of poles is always many times higher with these machines than with turbo alternators. The rotor diameters

Figure 10.3 96 kVA alternator rotor with protruding poles.

of these alternators reach up to 20 m and their length reaches several meters. These alternators, powered by water turbines, referred to as hydro alternators, are used in *hydroelectric power plants.*

Figure 10.3 shows the alternator rotor with excellent poles having a power of 96 kVA.

The most common high-performance HV electric motors are *three-phase asynchronous electric motors* with a *short-circuit armature* and a ring armature.

Asynchronous motors with a short-circuit armature consist of a stator, the winding of which is supplied with three-phase alternating current, which generates a rotating magnetic field in it. In this field, there is a rotor, formed by a ferromagnetic core, in which a cage winding is placed, formed by interconnected aluminum or copper conductors. The action of the rotating magnetic field induces an alternating current in these conductors, generating an alternating magnetic field. The mutual force of both of these fields creates the engine torque.

Figure 10.4 shows a section of the rotor of a three-phase asynchronous electric motor with a short-circuit armature.

An *asynchronous motor with a ring armature* has a rotor winding made of insulated conductors, which are placed in the grooves of its ferromagnetic core. These windings are supplied with alternating three-phase current via slip rings, which is connected either in a star or in a triangle. Three carbon brushes rest on the rings, to which the control rotor circuit can be connected. Higher power engines tend to have brush dippers with a ring short.

10.2 Properties of HV Rotary Electrical Machines

The basic characteristics of all rotating electrical machines include their electrical input, power, efficiency, power factor, speed, or chute.

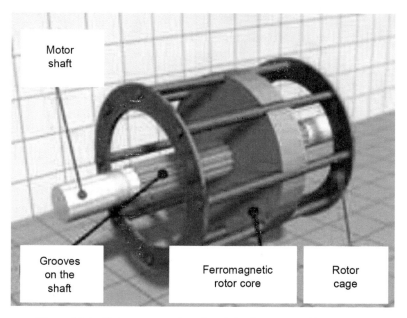

Figure 10.4 Rotor cross section of an induction motor with armature.

The *efficiency of generators* is determined by the ratio of their electrical power to their mechanical input

$$\eta_g = \frac{P_e}{P_m}. \tag{10.1}$$

The *efficiency of motors* is determined by the ratio of their mechanical power to their active electrical input

$$\eta_m = \frac{P_m}{P_e}. \tag{10.2}$$

The efficiency of motors is also expressed by their *power factor*, defined by the ratio of active power to their apparent power

$$\cos \varphi = \frac{P_m}{S}. \tag{10.3}$$

The efficiency of electrical machines generally increases with their performance. The differences between these powers and wattages are caused by their mechanical and electrical losses.

Mechanical losses include, in particular, friction in the bearings and aerodynamic losses caused by rotor rotation. *Electrical losses* are caused by losses in the ferromagnetic circuits of machines and losses caused by the passage of alternating current through their windings. Both types of losses cause the machines to heat up, and, therefore, these machines must be cooled, most often with air or compressed nitrogen.

For example, a generator with a nominal power of 1.8 MVA achieves an efficiency of 96.1%, while its mechanical losses are 8 kW, i.e., 0.45%, losses in magnetic circuits are 22.7 kW, i.e., 1.26%, losses in the windings stator rotor are 48.6 kW, i.e. 2.14%, and additional losses are 9 kW, i.e., 0.5%. An asynchronous motor with a short-circuit armature having a nominal power of 440 kVA achieves an efficiency of 95.5%, while its mechanical losses are 2.3 kW, i.e., 0.5%, losses in magnetic circuits are 4.2 kW, i.e., 0.9%, the losses in the stator and rotor windings are 10 kW, i.e., 2.1%, and the additional losses are 4.6 kW, i.e., 1%.

The *electric generator determines the frequency of the generated alternating current* by its *synchronous speed* n_s and the number of the stator pole p

$$f = \frac{1}{n_s \cdot p}. \tag{10.4}$$

If the frequency of the generated alternating current is to be 50 Hz, then the rotor of the single-pole generator has a rated speed of 3000 rpm. For a two-pole machine, the rated speed is 1500 rpm. Hydropower generators have a significantly higher number of rotor polesnes. For example, the 64-pole rotor has a rated speed of 93.75 rpm.

The speed of asynchronous motors is always lower than the synchronous speed n_s. The difference between these speeds defines their *slip*

$$s = \frac{n_s - n}{n_s}. \tag{10.5}$$

At steady state, the torque of asynchronous armature motors is briefly dependent on their *maximum torque* M_m and slip

$$M = \frac{2 \cdot M_m}{\dfrac{s}{s_z} + \dfrac{s_z}{s}}, \tag{10.6}$$

where s_z is the *reversal slip* at which the torsion of the induction motor reaches its maximum value. For large armature asynchronous motors, the slip reaches a few percentages, and, therefore, their torque is approximately

$$M \approx 2 \cdot s \cdot \frac{M_m}{s_z}. \tag{10.7}$$

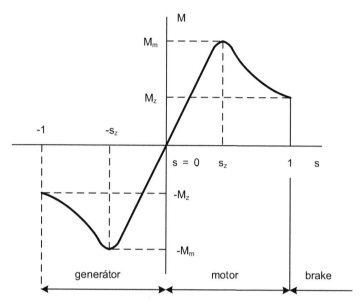

Figure 10.5   Torque characteristic of a short-circuit armature induction motor.

where M_m is the maximum torque moment and M_z is the traction torque moment and area of its use.

Figure 10.5 shows the *torque characteristic* of an asynchronous motor with a short armature with the marked *maximum torque M_m*, the *engagement toque M_z*, and the area of its use.

10.3 Diagnostics of HV Rotary Electrical Machines

Diagnosis of HV properties of rotary electrical machines is based mainly on standards EN 60034-1+A+A2, ed. 2 *Rotating electrical machines – Part 1: Nominal data and properties*, EN 60034-9+A1, *Rotating electrical machines, noise limits*, IEC 93 (34 6460) *Tests of rigid electrical insulating materials*, and EN 60 270 *High voltage test technique, measurement of partial discharges*. Diagnosis of large rotary electrical machines is performed either in offline or online mode.

10.3.1 Offline diagnostics of HV rotary electrical machines

Standard *online diagnostic tests* of rotary electrical machines include voltage tests of DC, AC, and pulse voltages, measurement of insulation resistance and its voltage dependence, including determination of polarization index of stator and rotor winding insulation system, measurement of frequency dependence of winding transmission impedance, and measurement of partial discharges in these systems by galvanic method and acoustic probe. Based on many years of measurements and experience with diagnostics of large HV electric rotating machines, the evaluation criteria of these tests were determined (see Table 10.1).

10.3.2 Online diagnostics of HV rotary electrical machines

Typical *online diagnostic methods* of rotary electrical machines include measurement of ozone detection in air and hydrogen purity in cooling systems of HV machines, frequency analysis of current and scattering magnetic field, analysis of their starting currents, and measurement of noise and vibration.

10.3.2.1 Ozone detection measurement

The presence of *ozone*, as a strong oxidant, reduces the life of insulation systems of electrical machines. Ozone is formed during partial discharges in these systems. Either *gas chromatography methods* or *photometric* and *colorimetric methods* are used to determine the ozone concentration.

The *photometric method* is based on the ability of ozone to absorb UV radiation in the 254-nm wavelength band. The ozone concentration in this analysis is determined by the difference between the absorption of UV radiation in the measured air sample and the absorption of this radiation in the reference air sample in the absence of ozone.

The *calorimetric method* is based on the absorption and reaction of ozone in a chloroform indicator solution, during which the blue indigo solution decolorizes. The change in its color is evaluated by means of a calibration curve, which determines the concentration of ozone in the air.

The reference volume concentration of ozone in the air is considered to be a volume concentration of less than 0.01 ppm, i.e., less than 10^{-8}. An acceptable volume concentration of ozone in the cooling air system of HV electrical machines is considered to be less than 1 ppm, i.e., 10^{-6}.

10.3.2.2 Measurement of hydrogen purity in cooling systems of electrical machines

The *purity of hydrogen* in the cooling systems of HV electric rotary machines is affected by substances that are excreted by the thermal stress of their solid insulators. These are mainly methane (CH_4), ethane (C_2H_6), ethylene (C_2H_4), acetylene (C_2H_2), carbon monoxide (CO), and carbon dioxide (CO_2). Similar to ozone measurement, gas chromatography is used to determine their concentrations.

Table 10.1   Evaluation criteria for testing large HV electrical machines.

Method used	Measured parameter	Criteria evaluation
Test of alternating voltage of stator and rotor winding	Breakthrough	Satisfactory if no breakthrough occurs
Stator voltage test with DC voltage	Insulation current	Suitable if the current does not increase or does not jump
Stator winding test with 0.1-Hz AC voltage	Jump	Suitable if there is no jump or break
Stator winding thread insulation test by SFRA frequency analysis	Stator insulation of stator winding	It is suitable if there is no significant difference between the course of the frequency transmission of the previously impedance and the simultaneously diagnosed isolation
Stator winding resistance and polarization index	10-minute polarization index P_{r600}	Suitable for $P_{r600} \leq P_{r60}$
Measurement of voltage dependence of insulation resistance with DC voltage	Insulation resistance R_{iz600}	On all voltage coils, the insulation resistance must increase with increasing voltage
Measurement of insulation resistance of rotor winding	Insulation resistance R_{iz60}	For $R_{iz60} \geq 0.5$ MΩ suits, for $R_{iz60} = 0.5$–0.1 MΩ satisfies the condition of more frequent measurements, for $R_{iz60} < 0.1$ MΩ does not suit
Measurement of polarization index of rotor winding	Polarization 1-minute index P_{i60}	For $P_{i60} \geq 1$ suits, for $P_{i60} < 1$ not suit
Measurement of capacitance and loss factor $tg\delta$ of stator winding	Loss factor $tg\delta$	New machine: $tg\delta \leq 0.035$, machine in operation: $tg\delta(U_n) \leq 0.085$, $tg\delta(0.2 \cdot U_n) \leq 0.035$, $tg\delta(0.6U_n - tg\delta(0.2 \cdot U_n) \leq 0.01$, $tg\delta/DU_n \geq 0$
Stator winding time constant measurement	Time constant τ	Time constant change $D\tau \geq 0$
Measurement of partial discharges of the stator winding by galvanic method	Apparent charge q	New machine: $q \leq 10$ nC; machine in operation: 10–30 nC, it is recommended to measure during operation; if it is still $q > 30$ nC, the machine does not comply
Measurement of partial discharges of the stator winding by an acoustic probe	Acoustic probe	Localization method to determine the position of partial discharges
Measurement of partial discharges of stator windings by differential electromagnetic probe	The difference in the level of partial discharges in the individual stator grooves	When the ratio of the actual value of the discharges to its mean values > 5, the properties of the insulation system in its respective groove have changed

10.3.2.3 Frequency analysis of currents and scattering magnetic fields

This method is intended for *online diagnostics* of asynchronous motors with short-circuit armature and/or synchronous motors with asynchronous starting. Because geometric inaccuracies in their rotors and stators can occur in each of these machines, interfering alternating magnetic fields are created in their air gaps. The speed of rotation of these fields is always different from the speed of rotation of the majority magnetic field of the diagnosed motors. The resulting asymmetries of the magnetic fields cause their successive, backward, and zero components.

In addition, frequency components proportional to its speed appear in the time course of the stator current of the motor and in its scattering magnetic field. Their occurrence, proportional to the asymmetry of the magnetic fields, is caused by faulty rotor installation and its mechanical and electrical faults.

The measurement of time courses and their corresponding frequency characteristics of the supply currents of stators of large HV electrical machines is performed on the secondary side of current transformers on one or more phases. During these measurements, the time course of the secondary current of the measuring current transformer is digitized and its frequency spectrum is determined by means of a *discrete Fourier transform.*

Subsequently, the frequencies of the sidebands of this spectrum are determined

$$f_1 = f_s - 2 \cdot f_s \cdot s \qquad\qquad f_2 = f_s + 2 \cdot f_s \cdot s, \qquad\text{(10.8a, b)}$$

where f_s is the frequency of the supply current and

$$s = \frac{n_s - n}{n_s} \qquad\qquad\text{(10.9)}$$

is the *slip* of the motor at the moment of its measurement.

The *speed* is determined by the number of revolutions per minute

$$f_{ot} = \frac{n}{60}. \qquad\qquad\text{(10.10)}$$

If the engine speed is not known at the time of measurement, the frequencies f_{x1} and f_{x2} shall be determined from its rated speed.

Figure 10.6 shows the time courses of the stator current of an asynchronous motor with a short armature with an intact and broken rotor cage.

The measurement of the *frequency characteristics* of the scattering field of motors is performed at their rated load and exposed ferromagnetic covers using ferromagnetic probes.

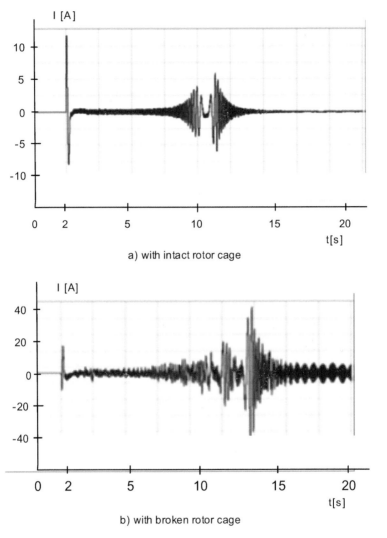

Figure 10.6 Time courses of the stator current of an induction motor with a short-circuit armature.

Figure 10.7 shows the *amplitude frequency spectra* of the stator current *I* and the intensity of the scattering magnetic field *H* of an asynchronous motor with a short-circuit armature with a broken rotor cage.

When evaluating the frequency spectra of stator currents and the scattering magnetic field intensities of asynchronous motors with a short armature, the difference between the amplitudes of the fundamental harmonic

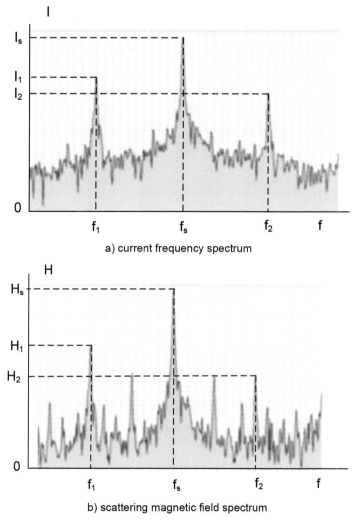

a) current frequency spectrum

b) scattering magnetic field spectrum

Figure 10.7 Frequency spectra of strator current and scattering magnetic field of an asynchronous motor with a short-circuit armature with a broken rotor cage.

components of these quantities at frequency f_s and the amplitudes of these components corresponding to frequencies f_1 and f_2 is determined.

Figure 10.8 shows the time course of the stator current I and its corresponding frequency spectrum of an asynchronous motor with a short armature with an *eccentric bearing* of its rotor, which is caused by the *bending* of its shaft.

Figure 10.9 shows the time courses of the stator current of an induction motor with a short armature with an undamaged and damaged stator winding.

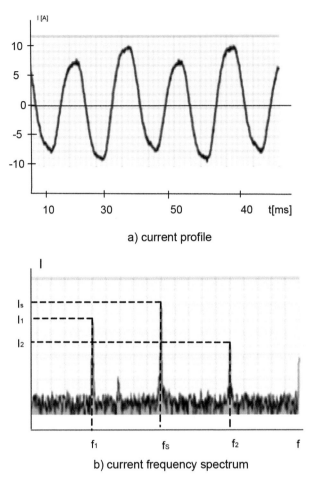

a) current profile

b) current frequency spectrum

Figure 10.8 Time course of the stator current and its corresponding frequency spectrum of an asynchronous motor with a short armature with eccentric mounting of its rotor.

10.3.2.4 Measurement of machine noise

The noise source of electric rotating machines is electromagnetic, mechanical, and ventilation noise.

The cause of the *electromagnetic noise* is the vibration of the machine frame, or its other parts, which is caused by the electromagnetic forces of its magnetic circuit. The frequency spectrum of this noise is discrete with frequencies that are multiples of the frequency of the motor supply current. *Mechanical noise* is caused by bearings and dynamic unbalance of rotating machine parts. *Ventilation noise* is generated by the operation of fans of flowing gases in the cooling system of machines.

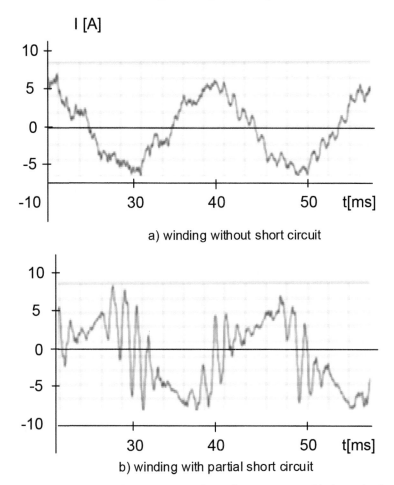

Figure 10.9   Time courses of stator currents of asynchronous motor with short -circuit armature with undamaged and damaged stator winding.

Noise diagnostics of rotary electrical machines is based on comparing the acoustic air pressure and sound frequency spectra of these machines with the acoustic pressures and frequency spectra of new or newly refurbished machines. Therefore, the time change of acoustic pressures and their frequency spectra is monitored. Noise measurements are performed at rated speed and rated machine load.

Acoustic pressure is caused by dilution and densification of the material environment, in which noise propagates in the form of acoustic waves. In technical practice, the sound pressure level is used to express the noise level, which is defined by the logarithmic ratio of the variable sound pressure

Table 10.2   Sound pressure level correction by octave filter.

f_m [Hz]	63	125	250	500	1000	2000	4000	8000
k_{Ai} [dB]	−26.2	−16.1	−8.6	−3.2	0	−1.2	−1.0	−1.1

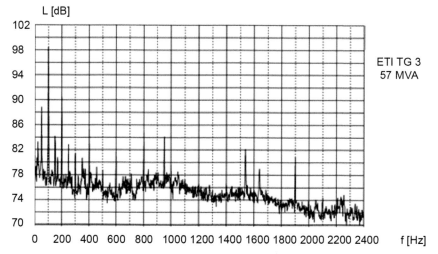

Figure 10.10   Frequency dependence of the sound pressure level of the turbogenerator.

component to the *reference pressure* $p_0 = 2 \cdot 10^{-5}$ [N/m²], defining the *audibility threshold*

$$L = 20\log \frac{p}{p_0}, \tag{10.11}$$

Noise meters, the main parts of which are *piezoelectric* or *condenser micro-phones*, are used to measure noise. In practice, a *modified sound pressure level* is used by the octave filter with a medium frequency f_m, which suppresses the measured pressure values in the individual *octave bands*

$$L_A = 10\log \sum_{i=1}^{n} 10^{\frac{L_i + k_{Ai}}{10}}, \tag{10.12}$$

where L_i is the sound pressure level in the *i*th *octave band* in dB and k_{Ai} is the correction in the respective octave band in dB. Table 10.2 shows these corrections for the center frequencies of the octave filter frequency bands.

Figure 10.10 shows an example of the frequency spectrum of the sound pressure level of an unloaded turbo alternator with a power of 57 MVA, from which the majority frequency component of the sound pressure of noise at a frequency of 100 Hz, reaching the level of 98 dB, is evident.

10.3.2.5 Online monitoring of partial discharges

Either galvanic methods or current transformers are used for online monitoring of partial discharges of rotary electrical machines.

High-frequency bushing current transformers with a ferrite core are usually connected to the supply wires of machines. The frequency range of these transformers is usually 100 kHz to 30 MHz at a rated load of 50 Ω. This allows them to be connected to partial discharge meters by conventional coaxial cables of the same wave impedance in order to eliminate the reflection of pulse signals which are fed to them by these cables.

In large and particularly important rotary electrical machines, such as turbogenerators, *electrostatic* or *electromagnetic probes* are used for online detection of partial discharges, which are built directly into the grooves of the stator windings of these machines. *Ultrasonic* and *high-frequency partial discharge sensors* are used for online detection outside rotary machines.

10.3.3 Shaft voltages and currents

Shaft stresses and corresponding shaft currents, caused by the asymmetry of the magnetic circuits of rotary machines and their excitation by *non-harmonic currents*, are the cause of increased wear of the bearings, which is also called their *electro erosion*. Therefore, the identification and measurement of these quantities is essential in the diagnostics of rotary electrical machines in order to increase the length of their trouble-free operation.

10.3.3.1 Generation of shaft voltages and currents

The cause of shaft voltages of rotating electrical machines, excited by alternating currents, is geometric inaccuracies of stators and rotors, or their axial misalignment.

In addition, the DC magnetization of their ferromagnetic shafts is manifested in DC machines. When the magnetic flux $d\Phi/dt$ of rotary electrical machines changes over time, which is caused by geometric inaccuracies in the production of their stators and rotors, a shaft voltage is generated.

$$u_h(t) = \frac{d\Phi}{dt}. \tag{10.13}$$

The frequency of alternating shaft voltages is given by the ratio of the number of poles p of the rotor to the number of gaps m in the magnetic circuit of the stator

$$f_i = \frac{p}{m} f_s \tag{10.14}$$

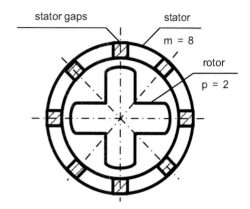

Figure 10.11   Rotor and stator of an
AC electric machine.

where f_s is the frequency of the excitation alternating current of the machine (50 Hz).

Figure 10.11 shows the magnetic circuit of the stator of an AC machine with eight gaps and a rotor with two pairs of poles.

Shaft voltages of large electrical machines can reach units up to tens of volts. If the shaft voltages exceed the insulating strength of the oil layer of ball or roller bearings, which is several volts, then the bearings of the machines go through *bearing currents*, which can reach up to tens of amperes at large rotations.

In addition, the excitation of rotary machines by non-harmonic excitation currents from the frequency converters causes disturbing current pulses, which are caused by the asynchronous switching of the power switches of the frequency converters. These pulses are transmitted through parasitic capacitances between the rotors and stators of AC machines, which in large rotary electrical machines reach tens to hundreds of nanofarads. The current pulses then pass through the bearings of the machines and thus cause their *electro erosion*. In large electrical machines, these pulses can reach up to tens of amperes.

10.3.3.2 Measurement of shaft voltages and currents

The *shaft voltages* of electrical machines are measured between collectors (brushes) located at both ends of the shafts, one of which is grounded and the other is ungrounded.

To *measure shaft currents*, *toroidal current transformers* are usually used, threaded on the machine shafts between the grounded collector and the machine, or these transformers are threaded on the ground conductor of the grounded collector. The prerequisite for achieving the relevant results is that

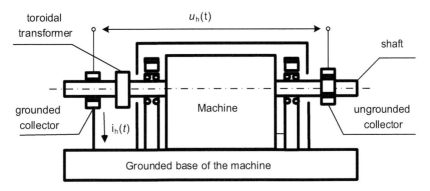

Figure 10.12 Measurement of shaft voltages and currents.

the insulation strength of the bearing lubricant is greater than the maximum values of the measured shaft stress (Figure 10.12).

If the shaft current i_h passes through the toroidal transformer bushing, on the ferrite core of which a secondary winding with N turns is wound, then for the secondary current of the transformer i_2 and the output current of the converter, the voltage current with the inverting operational amplifier applies

$$u_2 = -2R_2 \cdot i_2 = -2 \cdot R_2 \frac{i_h}{N}. \tag{10.15}$$

A *Hall probe* is also used to measure shaft currents, which measures the tangential component of induction B on a shaft of radius r

$$B_t = \mu_0 \cdot \mu_r \frac{i_h}{2 \cdot \pi \cdot r}, \tag{10.16}$$

where $\mu_0 = 4\pi \cdot 10^{-7}$ [Hm^{-1}] is the *permeability of the vacuum* and μ_r is the *relative permeability* of the shaft material, which is approximately $8 \cdot 10^3$ for steel shafts. The output voltage of the Hall probe is then determined by the product of this induction with the Hall probe constant k_H with its excitation current I_b

$$u_H = k_H \cdot I_B \cdot B_t = k_H \cdot I_B \cdot \mu_0 \cdot \mu_r \frac{i_h}{2 \cdot \pi \cdot r}. \tag{10.17}$$

Figure 10.13 shows the possibilities of measuring shaft currents with a toroidal current transformer and a Hall probe.

If the value of the shaft voltage approaches the electrical strength of the bearing lubricant, which is several volts, then *partial discharges* occur in this layer of lubricant, the mean value of which reduces the measured data

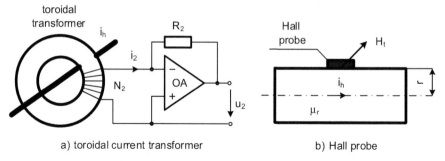

a) toroidal current transformer   b) Hall probe

Figure 10.13   Measurement of shaft currents.

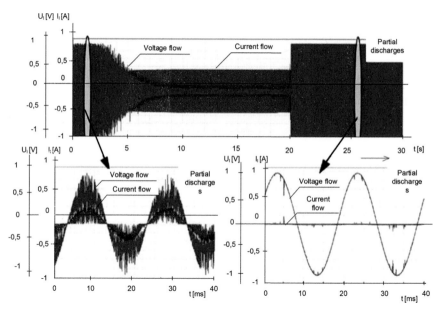

Figure 10.14   Time courses of bearing voltages and currents when stopping and starting the motor.

of shaft currents. A specific feature of bearing lubricants is the spontaneous restoration of their electrical strength after a longer period of electrical stress.

Figure 10.14 shows the changes in the insulating properties of the bearing when stopping and restarting a 100-kVA motor, including the representation of partial discharges.

Figure 10.15 shows the time changes of the insulating properties of the bearing lubricant during degradation and spontaneous restoration of its electrical strength.

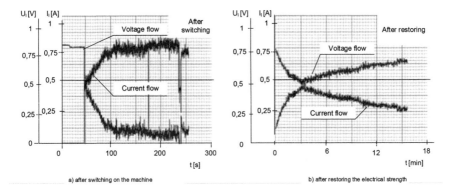

Figure 10.15 Insulation properties of bearing lubricant.

ceramic bearing insulation

Figure 10.16 Hybrid insulated bearing.

10.3.3.3 Means of suppressing shaft voltages and currents

Based on a long-term evaluation of failures of rotating electrical machines, it was found that bearing currents cause up to 50% probability of occurrence of all failures of these machines.

In addition to their suitable design, insulated bearings and earthling brushes are used to suppress the occurrence and action of bearing voltages and currents of large electric rotary machines, or *short-circuited threads* are mounted on their shafts during their production.

Insulated bearings, also called *hybrid bearings*, are provided on their outer surface with a ceramic layer that has a similar thermal expansion as the shaft material. The insulating strength of this layer is several hundred volts and its capacity varies according to the size and type of bearings in the range of hundreds of pF to nF units (Figure 10.16).

In the case of a simple machine with two bearings, it is sufficient to insulate only one of the bearings. Shaft sets, consisting of several electrical machines, should be electrically insulated with their connectors.

The second option is to lead the shaft currents to ground using grounding brushes. Brushes are usually made of electro graphite, metal graphite, or metal microfiber materials. *Grounding brushes* are usually located on the inside of the machine's bearing shields to prevent contamination. The

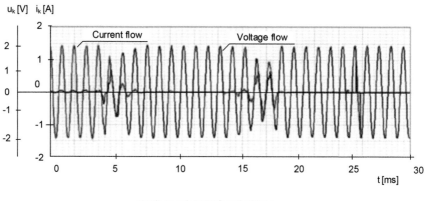

a) voltage and current flows for 30 ms

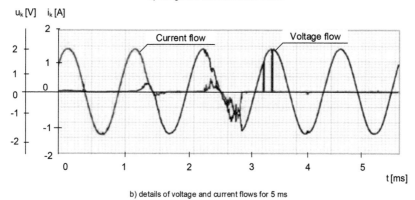

b) details of voltage and current flows for 5 ms

Figure 10.17   Time courses of voltages and currents passing through the collecting brush.

disadvantage of using collection brushes is the occurrence of transient stresses on them and their limited service life.

Figure 10.17 shows the time courses of voltages and currents passing through the collecting brush when the motor is started.

Figure 10.18 shows the transient characteristics of voltages and currents passing through a metal graphite and microfiber brush.

To assess the suitability of the use of brushes in terms of the efficiency of current pulses with a high content of high-frequency components, it is useful to know the frequency dependence of the V–A characteristics of these brushes.

Figure 10.19 shows the frequency dependences of the V–A characteristics of the most used types of collection brushes.

It follows from the above frequency V–A characteristics that both types of brushes increase in impedance with increasing frequency, while the metal

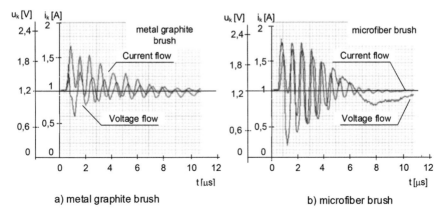

Figure 10.18 Transient characteristics of brush voltages and currents.

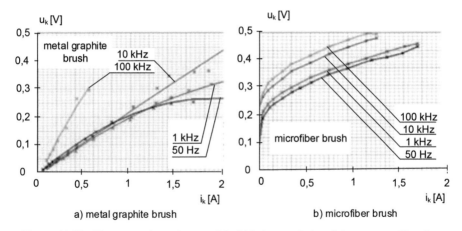

Figure 10.19 Frequency dependences of the V-I characteristics of the most used brushes.

impedance of graphite brushes is lower at lower frequencies compared to the impedances of microfiber brushes. However, the disadvantage of graphite brushes is their higher wear and the related need for more frequent replacement.

References

[1] Hauschild, W., Lemke, E. High Voltage Test and Measuring Techniques. Springer Verlag 2018. ISBN: 978-3-319-97459-0.
[2] Záliš, K. *Partial discharges in the insulation systems of electrical machines.* Academia, publishing house of the Academy of Sciences of the Czech Republic, Prague 2005. ISBN 80-200-1358-X.

[3] Mentlík V., Pihera, J., Polanský|, R., Prosr, P., Trnka. P. *Diagnostics of electrical equipment.* BEN-technical literature, Prague 2008. ISBN 978-80-7300-232-9.

[4] Martínek, P.: *Theoretical and experimental analysis of partial discharges in dielectric of electric machines.* Dissertation thesis, University of West Bohemia, Pilsen 2005.

[5] Klasna, J., Martínek, P., Fanta, R., Pihera, J. *Partial discharges in the high-voltage insulation of turbo-generators.* International Conference Electric Power Engineering. EPE 2010.

[6] Hubáček, J. *Partial discharges on generators of large power and their measurement.* Bachelor Thesis, University of West Bohemia, Pilsen 2011.

[7] Hubáček, J. *Unconventional methods of monitoring partial discharges.* Bachelor Thesis of University in Pilsen 2013.

[8] *High voltage test technology – measurement of partial discharges*, EN 60270.

[9] Humlová, V. *Acoustic responses of partial discharges in the diagnosis of insulation defects.* Thesis. University in Pilsen, 2019.

[10] Sill, U., Zormner, W. *Steam Turbine Generators Process Control and Diagnostics.* Siemens 1996. ISBN: 13-978-3895780400.

[11] Boltze, M., Coenen, S., Tenbohlen, S. *Prospects and Limits of on-site PD Measurement.* TU Stuttgart, Germany 2011.

[12] Muhr, M. *Developments in Diagnosis of High Voltage Apparatus.* University of Technology Graz 2009.

[13] Muhr, M. *IEC 62478 A Prospective Standard for Acoustic and Electromagnetic Partial Discharge Measurements.* University of Technology Graz 2009.

[14] Lemke, E. etc. *Guide for Electrical Partial Discharge Measurements in compliance to IEC 60270.*

[15] EN 61 063 *Measurement of airborne noise emitted by steam turbines and their driven machines.*1998.

11

Diagnostics of HV Power Cables

Power HV cables to ether with generators and transformers form the basic system of energy networks. Therefore, diagnostics of their electrical and climatic properties is essential in increasing the reliability of these networks. The diagnostics of the own HV and VHV cables also includes the diagnostics of their connecting cable terminals.

11.1 Types of MV Power Cables

Single-wire or *multi-wire cables* with one or more aluminum or copper cores are used in the power system to distribute electricity.

PILC cables with *oil-impregnated paper* insulation were previously used in cable construction. Due to the homogeneous distribution of the electrostatic field on their insulating surface, perforated aluminum strips were placed on them. *XLPE cables*, which are *insulated with cross-linked polyethylene*, have been used to distribute electricity since the 1970s.

Single-conductor HV cables have electrical conductors consisting of several stranded copper or aluminum conductors 1, on which an inner conductive layer 2 is applied. To ensure the homogeneity of the electrostatic field, a conductive carbon layer 4 is used on the surface of the insulation. A textile tape 5 is wound on this layer, on which a shielding spiral with copper strips 6 is wound, polyethylene (Figure 11.1).

In this design, *single-conductor HV cables* are manufactured in the range of operating voltages of 6 kV, up to 18 kV with a cross-section of copper stranded cores in the range of 50–500 mm² with current carrying capacity in air of 240–927 A, and in the ground in the range of 220–750 A.

Multi-conductor cables usually consist of three or four separate conductors (Figure 11.2).

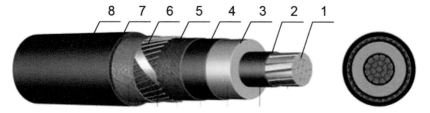

a) insulating layers of the cable   b) cross section of the cable

Figure 11.1   Single conductor HV cable.

a) three-core cable   b) four-core cable

Figure 11.2   Multi-conductor HV cables.

Figure 11.3   Three-conductor HV cable with cooling ducts.

The composition of the insulation of these cables is similar to the composition of the insulation of single-conductor cables, with the difference that some cables have, in addition to the outer shielding, also the inner shielding of its individual conductors.

Special multi-conductor cables are cables cooled by *compressed nitrogen*. Figure 11.3 shows a cross-section of a three-core submarine cable with an operating voltage of 600 kV, which can transmit power up to 2.2 GVA. The cable is equipped with cooling channels through which compressed nitrogen flows.

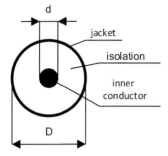

Figure 11.4 Coaxial cable cross section.

In addition, these cables are equipped with reinforced insulation against mechanical damage and aggressive salt water.

11.2 Properties of High-voltage Power Cables

In addition to electrical strength, the basic electrical properties of power HV cables also include their throughput and leakage resistance, capacity, inductance, power dissipation, and wave impedance.

The *cross-section* of a single-wire coaxial power cable is shown in Figure 11.4.

The through resistance of a round conductor per unit of its length d and cross-section S is determined by the specific resistance of the metal ρ from which the conductor is made

$$R_p = \frac{\rho}{S} = \frac{4 \cdot \rho}{\pi \cdot d^2} \left[\Omega \cdot m^{-1} \right].$$

$$(11.1)$$

For economic reasons, the so-called *electrical aluminum* with 95% aluminum content is used for the production of common HV and VHV cables, the specific resistance of which is $\rho_{Al} = 0.0285 \cdot 10^{-6}$ [Ωm]. For more demanding applications, the so-called *conductive copper* with 95.95% copper content is used for the production of cables, whose specific resistance is $\rho_{Al} = 0.0285 \cdot 10^{-6}$ [Ωm].

If an alternating current of frequency f passes through the cable conductor, then due to the *skin effect*, the current density increases toward its surface, which causes its resistance to increase to the value

$$R_p(f) = R \cdot \frac{d}{2} \sqrt{\frac{\mu \cdot f}{\rho}} \left[\Omega \cdot m^{-1} \right],$$

$$(11.2)$$

where ρ is the resistivity of the conductor material and μ is its permeability.

For example, for a copper conductor with a diameter $d = 20$ mm and at a mains frequency $f = 50$ Hz, the ratio between the resistors $\delta R_p(f) = 1.005$, which is practically negligible.

If the cables carry non-harmonic current waveforms, caused, for example, by the pulse action of DC/AC converters of direct current to alternating current, then the increase in resistance $R_p(f)$ may no longer be negligible. For example, at a transmitted current frequency of 5 kHz, the resistance ratio $\delta R_p / R_p(f) = 1.05$.

The leakage resistance of a coaxial cable per unit of its length with the circular cross-section of the conductor d and the inner diameter D of its shielding sheath is determined by the specific volume resistance of the insulation ρ

$$R_{sv} = \frac{2 \cdot \pi \cdot \rho}{\ln \dfrac{D}{d}} \left[\Omega \cdot m^{-1} \right]. \tag{11.3}$$

The *capacity of a coaxial cable* per unit of its length relative to its grounded shielding sheath is determined by the permittivity of the insulation ε and the inner and outer diameters of the insulation d and D

$$C = \frac{2 \cdot \pi \cdot \varepsilon}{\ln \dfrac{D}{d}} \left[F \cdot m^{-1} \right], \tag{11.4}$$

where ε is the insulation permittivity, determined by the product of *vacuum permittivity* $\varepsilon_0 = 8.854 \cdot 10^{-12}$ [F·m⁻¹] with relative insulation permittivity ε_r, which is approximately 2.4 for cross-linked polyethylene.

The *inductance of a coaxial cable* per unit length is determined by the following equation:

$$L = \frac{\mu}{2 \cdot \pi} \cdot \ln \left(\frac{D}{d} \right) \left[H \cdot m^{-1} \right], \tag{11.5}$$

where μ is the permeability of the insulation, determined by the product of the *permeability of the vacuum* $\mu_0 = 4 \cdot \pi \cdot 10^{-7}$ [H·m⁻¹] with the relative permeability of the insulation μ_r, which is 0.95 for cross-linked polyethylene.

Based on these parameters, the electrical properties of the coaxial cable per unit of its length can be expressed by the substitution diagram in Figure 11.5.

The *wave impedance* of the cable is then defined by the ratio of the through impedance of the cable to its leakage admittance

$$Z_0 = \sqrt{\frac{R_p + j \cdot \omega \cdot L}{\dfrac{1}{R_{sv}} + j \cdot \omega \cdot C}} \; [\Omega], \tag{11.6}$$

where $\omega = 2 \cdot \pi \cdot f$ is the circular frequency of the alternating current.

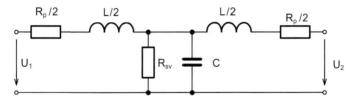

Figure 11.5 Replacement diagram of coaxial cable.

Since at the mains frequency $f = 50$ Hz $R_p << \omega \cdot L$ and $1/R_{sv} << \omega \cdot C$, it applies to the wave impedance of the cable

$$Z_0 \approx \sqrt{\frac{L}{C}}\,[\Omega].$$ (11.7)

Knowing the input impedance of the unloaded cable Z_n and the impedance of the shorted cable Z_k, its wave impedance can be determined by the square diameter of these impedances

$$Z_0 = \sqrt{Z_n \cdot Z_k}\,[\Omega].$$ (11.8)

The *pulse propagation* speed of the coaxial cable is determined by the following equation:

$$v = \frac{c}{\sqrt{\varepsilon_0 \cdot \varepsilon_r \cdot \mu_0 \cdot \mu_r}}\,[\text{ms}^{-1}],$$ (11.9)

where $c = 3 \cdot 10^8$ [ms^{-1}] is the speed of light propagation in vacuum, $\varepsilon_0 = 8.854 \cdot 10^{-12}$ [F·m^{-1}], and $\mu_0 = 4 \cdot \pi \cdot 10^{-7}$ [H·m^{-1}] are permittivity and vacuum permeability, and ε_r and μ_r are the relative permittivity and permeability of the cable insulation. Assuming that $R_p << \omega \cdot L$ and $1/R_{sv} << \omega \cdot C$, the pulse propagation rate in the cable

$$v \approx \frac{1}{\sqrt{L \cdot C}}\,[\text{ms}^{-1}].$$ (11.10)

For example, for HV coaxial cable with polyethylene insulation with a capacity of 200 nF/km and an inductance of 0.59 mH/km, the pulse propagation speed is $v = 1.34 \cdot 10^8$ [ms^{-1}]. If the coaxial cable is loaded with an impedance Z_2 different from the wave impedance Z_0, then when it is excited by the impulse signal, a reflection is created at its end, which is defined by the *reflection factor.*

$$p = \frac{Z_2 - Z_0}{Z_2 + Z_0}.$$ (11.11)

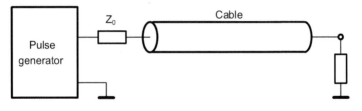

a) cable excitation

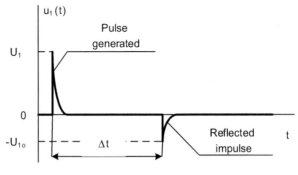

b) with the cable end disc

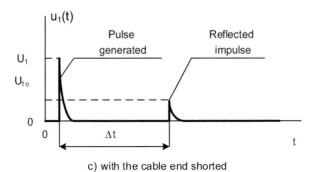

c) with the cable end shorted

Figure 11.6   Time pulses in the cable.

If $Z_2 = Z_0$, then $p = 1$ and the pulse reflection does not occur, at $Z_2 > Z_0 \, p > 0$, and at $Z_2 < Z_0, p < 0$.

If the cable is disconnected at its output, then the reflected pulse has the same polarity as the excitation pulse. Conversely, if the cable is shorted at its end, then the reflected pulse has the opposite polarity with respect to the excitation pulse.

The time courses of the pulses in the cable are shown in Figure 11.6.

For periods, $\Delta T = l/v$, where l is the cable length.

The *power lost* by the passage of current through the cable has an active and a reactive component.

The *active power loss* of a cable per unit of its length is determined by the sum of the active losses at its through and leakage resistance

$$P_k = R_p \cdot I^2 + \frac{1}{R_s}\left[\frac{U_1 + (U_1 - R_p \cdot I)}{2}\right]^2 = R_p \cdot I^2 + \frac{(2 \cdot U_1 - R_p \cdot I)^2}{4 \cdot R_s}[W], \quad (11.12)$$

where I is the RMS value of the current flowing through the cable.

The *reactive power loss* of a cable per unit of its length is determined by the sum of the reactive losses on its capacity and inductance.

The cable loss factor is then defined by the ratio of active and reactive power

$$tg\delta = \frac{P_k}{Q_k}. \quad (11.13)$$

For example, the HV single-conductor coaxial cable with polyethylene insulation for AC operating voltage of 72 kV with a core diameter of 19.6 mm and an outer diameter of *polyethylene insulation* of 60 mm has a through resistance of the *aluminum core* of 0.1 Ω/km. The leakage resistance of this cable is 1 GΩ/km, the capacity is 0.2 µF/km, and the inductance is 0.59 mH/km. When 100-A AC passes through this cable, it has an active power dissipation of 101 W and a reactive power of 1850 VAR, which determines its power dissipation of 0.056.

11.3 Measurement of Properties of HV Power Cables

The diagnostics of the properties of HV and VHV cables is based on the standard EN 61442 or older standards ČSN 34 7010 and ČSN 34 7405. According to these standards, the electrical resistance of HV cables, their through and leakage resistances, loss factor, the occurrence of partial discharges, and the place and extent of their possible failure are verified. These parameters should be determined not only during the production of cables but also during their further operation.

11.3.1 Measuring the resistance of cables

To verify the electrical resistance of HV and VHV in Chapter 8 are used. Standard test methods are AC and pulse *voltage test methods*. The use of DC test voltage in the diagnosis of XLPE cables with cross-linked polyethylene

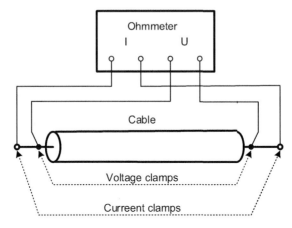

Figure 11.7   Measuring the through resistance of a cable by the four-wire method.

insulation is unsuitable because the DC space charge generated by using DC test voltage significantly deforms the electric field density of these cables, which increases the probability of cable defects.

The VLF (very low frequency) method is therefore used to verify the electrical strength of MV cables, in which the cable is tested with an alternating test voltage of 0.1–100 Hz with a maximum of three times its nominal value for 60 minutes.

11.3.2 Measuring the through resistance of cables

This measurement can be performed on *single-wire cables* if both ends are accessible. In the case of *multi-conductor cables*, one of their ends is always short-circuited, and the resistance of both of their conductors is measured at their other end.

In both cases, the through-resistance is always measured by a *four-wire method*, eliminating the measurement error caused by the resistance of the measuring cables (Figure 11.7).

In order to eliminate the possible occurrence of thermoelectric voltages at the voltage terminals of the measured cable, the measurement is performed for both polarities of the DC measuring current and the resistance of the cable conductor is determined by the mean value from both measurements.

If the measured resistance of the conductor differs from its nominal value by more than 5%, then it is possible to infer its defective production or damage during operation, e.g., *electro erosion*.

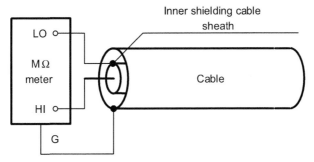

a) measuring the resistance between the inner conductor and the shielding sheath

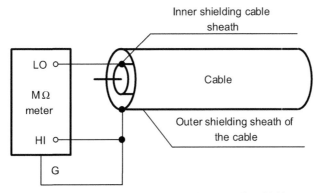

b) measuring the resistance between the shields

Figure 11.8   Measurement of cable leakage resistances.

11.3.3 Measurement of cable leakage resistances

For *single-conductor cables*, the leakage resistance between its core and the inner shield is measured. If the cable has internal and external shielding, the leakage resistance between these shields is also measured (Figure 11.8).

Because this is a measurement of large resistances, in the order of units up to hundreds of gigaohms, it is always necessary to ground the shielding sheaths of the measured cables. Voltage levels in the range of 2.5–5 kV are used to measure leakage resistances.

A typical HV cable should have a specific leakage resistance greater than 100 MΩ/km. If the measured leakage resistances of the cables differ from their nominal values by more than 5%, then it is possible to infer their defective production or damage during operation.

For *three-wire* and/or *four-wire cables*, not only the leakage resistances R_s between the conductors and the shield but also between the conductor

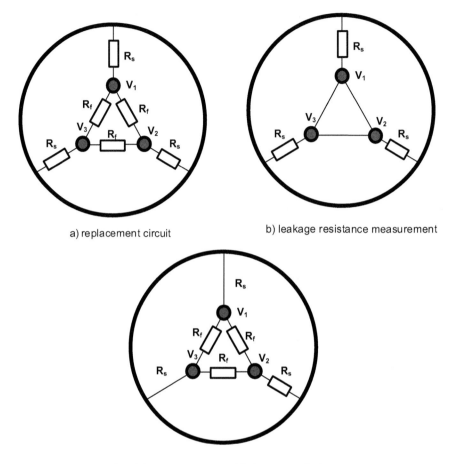

a) replacement circuit   b) leakage resistance measurement

c) resistance measurement between conductors

Figure 11.9   Measurement of leakage resistances of a three-wire cable.

resistances R_f between the individual conductors are measured. The measurement of these resistances is based on the substitute diagram in Figure 11.9.

The measurement of leakage resistances is carried out in two stages. In the *first stage* of the measurement, all three conductors of the cable are connected and the resistance between this connection and the shielding sheath of the cable is measured. Assuming that the leakage resistances between the conductors and the cable shield are identical, the measured resistance R_{m1} determines the leakage resistances

$$R_s = 3 \cdot R_{m1}. \tag{11.14}$$

In the *second stage* of the measurement, the two inner conductors of the cable are always connected to its shielding sheath and the resistance R_{m2} between this connection and the cable sheath is measured. Assuming that the conductor resistances R_f are the same, the resistance thus measured between the conductor resistances determines

$$R_f = \frac{6 \cdot R_{m1} \cdot R_{m2}}{3 \cdot R_{m1} - R_{m2}}.$$

The *total DC dielectric losses* of a three-conductor cable are determined by the sum of leakage losses s between the conductor losses in all phases

$$P_s = \sum_{i=1}^{3} \frac{U_s^2}{R_{si}} + \sum_{i=1}^{3} \frac{U_{fi}^2}{R_{fi}}, \qquad (11.15)$$

where U_{si}, $i = 1, 2, 3, \ldots$ are the *associated voltages* and U_{fi}, $i = 1, 2, 3, \ldots$ are the *phase voltages* of the three-phase network, for which $U_s = U_f \cdot \sqrt[3]{3}$ holds in a symmetrical arrangement.

11.3.4 Measurement of capacity and loss factor of cables

The procedure for measuring the capacity and loss factor of cables is similar to the measurement of their leakage resistances, while the *leakage loss impedances* Z_s and the *interconductor loss impedances* Z_f are determined. From these impedances, leakage and interconductor capacitances and loss factors are determined. The measurement of these loss impedances is based on the substitution scheme in Figure 11.10.

The measurement of loss impedances is again performed in two stages. In the *first stage* of the measurement, all three conductors of the cable are connected and the impedance between this connection and the shielding sheath of the cable is measured. Assuming that the capacitances and loss resistances between the conductors and the cable shield are identical, the real impedance component $\mathrm{Re}(Z_{m1})$ determines its *leakage loss resistances*.

$$R_{sz} = 3 \cdot \mathrm{Re}(Z_{m1}). \qquad (11.16)$$

Cable loss capacitances are determined by imaginary components of impedances $\mathrm{Im}\,(Z_{m1})$

$$C_{sz} = \frac{1}{3 \cdot \omega \cdot \mathrm{Im}(Z_{m1})}, \qquad (11.17)$$

In the *second stage* of the measurement, the two inner conductors of the cable are always connected to its shielding sheath and the impedance Z_{m2}

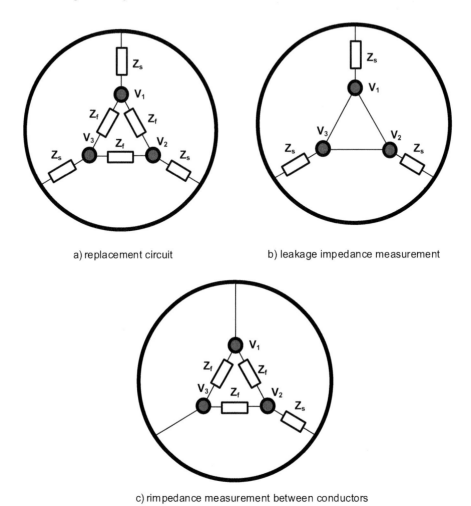

a) replacement circuit

b) leakage impedance measurement

c) rim pedance measurement between conductors

Figure 11.10   Measurement of leakage impedances of a three-wire cable.

between this connection and the cable sheath is measured. Assuming that
the conductor leakage resistances and capacitances are the same, the real
impedance components $\mathrm{Re}(Z_{m1})$ and $\mathrm{Re}(Z_{m2})$ determine the *interconductor
loss resistances*

$$R_{fz} = \frac{2 \cdot \mathrm{Re}(R_{m1}) \cdot \mathrm{Re}(R_{m2})}{\mathrm{Re}(R_{m1}) - \mathrm{Re}(R_{m2})}. \tag{11.18}$$

Interconductor loss capacities are determined by imaginary components of
impedances $\mathrm{Im}(Z_{m1})$ and $\mathrm{Im}(Z_{m2})$

$$C_{fz} = \frac{1}{2 \cdot \omega} \left[\frac{1}{\text{Im}(Z_{m2})} - \frac{1}{\text{Im}(Z_{m1})} \right]. \tag{11.19}$$

For *three-wire cables*, the so-called *effective loss capacity* is further defined

$$C_{efz} = \frac{1}{\omega} \left(\frac{3}{2 \cdot \text{Im}(Z_{m2})} - \frac{1}{6 \cdot \text{Im}(Z_{m1})} \right). \tag{11.20}$$

With the knowledge of resistance capacitances and capacitances, it is possible to determine the *leakage* and *leakage impedance loss factors*

$$tg\delta_{sz} = \omega \cdot R_{sz} \cdot C_{sz}, \qquad tg\delta_{fz} = \omega \cdot R_{fz} \cdot C_{fz}. \tag{11.21}$$

The *total AC dielectric losses* of a three-wire cable are determined by the sum of the AC leakage losses and between the AC conductor losses. These losses have a real and imaginary component.

The *real component of AC dielectric losses* determines the active component of these losses

$$P_s = \sum_{i=1}^{3} \frac{U_{si}^2}{R_{sz}} + \sum_{i=1}^{3} \frac{U_{fi}^2}{R_{fz}} [W], \tag{11.22}$$

which are usually substantially higher than the direct current dielectric losses caused by the leakage resistances of the cable.

The *imaginary component of AC dielectric losses* determines the reactive component of these losses

$$Q_s = \sum_{i=1}^{3} \frac{U_s^2}{\omega \cdot C_{sz}} + \sum_{i=1}^{3} \frac{U_{fi}^2}{\omega \cdot C_{fz}} [VAR], \tag{11.23}$$

which degrade the transmission *power factor* of cables

$$\cos\varphi = \frac{P_s}{Q_s} \tag{11.24}$$

Because the reactive component of the dielectric losses of HV cables loads the cables with current without participating in the transmission of active power, the reactive losses in the distribution cable networks are compensated by power factor compensators. These are formed by *compensating chokes*, the inductance of which is chosen so that the reactive losses of the capacitive character of the cables are compensated by the reactive losses of the compensating chokes.

Table 11.1 Recommended values of HV cable loss factor.

Cable insulation	Test voltage $2 \cdot U_n$	Test voltage $2 \cdot U_0 - U_0$	Cable
PILC	$< 20 \cdot 10^{-3}$	$< 5 \cdot 10^{-3}$	Serviceable cable
(impregnated	$20 \cdot 10^{-3}$ to $25 \cdot 10^{-3}$	$5 \cdot 10^{-3}$ to $7 \cdot 10^{-3}$	Slightly worsened condition
paper)	$> 25 \cdot 10^{-3}$	$> 15 \cdot 10^{-3}$	Risk of failure
XLPE	$< 6 \cdot 10^{-3}$	$< 10^{-3}$	Serviceable cable
(cross-linked	$6 \cdot 10^{-3}$ to $8 \cdot 10^{-3}$	10^{-3} to $2 \cdot 10^{-3}$	Slightly worsened condition
polyethylene)	$> 8 \cdot 10^{-3}$	$> 2 \cdot 10^{-3}$	Risk of failure

Complete compensation of reactive cable losses is conditioned by the inductance of the chokes

$$L = \frac{1}{4 \cdot \pi^2 \cdot f^2 \cdot C_{efz}}. \tag{11.25}$$

The measurement of the capacity and loss factor of MV cables is carried out in the range of up to twice the nominal voltage U_n of the cable at frequencies of 0.1–10 Hz.

Recommended values of loss factor of cables with impregnated paper and polyethylene insulation are given in Table 11.1.

Ideally, the measured values of the loss factor should not increase significantly with increasing voltage and frequency of test voltages.

For example, Baur, a manufacturer of digital systems for measuring MV cables, recommends measuring their loss factors at three voltage levels of test voltages U_n, $1.5 . U_n$, and $2 . U_n$ and at frequencies of 0.1 and 100 Hz.

Before measuring, the cables must always be disconnected from all devices and their residual charges discharged. The measured loss factors shall be determined from the mean of eight consecutive measurements, the standard deviation being less than 5%.

Figure 11.11 shows the results of measuring the voltage and frequency dependences of the loss factors of individual interconductor insulations of a three-wire cable.

If the loss factors increase significantly with increasing test AC voltage, then this phenomenon is usually caused by wetting of the insulation and the subsequent formation of the so-called *tree partial discharges*. In the extreme case, these discharges can increase to the point of cable insulation.

A frequent cause of insulation penetration into the cable insulation is a breach of the outer insulation of the cable or its imperfect connection by means of MV terminals. If possible, the two fault sources must be separated when measuring the loss factors.

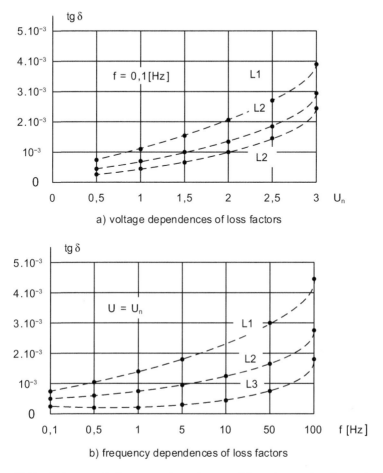

Figure 11.11 Voltage and frequency dependences of loss factors by three-wire cable insulation.

11.3.5 Measurement of partial discharges in cables

Due to their low charges, of the order of nC units per 1 km of their length, the *differential galvanic method* is used to measure partial discharges in HV cables (Figure 11.12).

This method requires the use of a two-channel partial discharge meter (PDM), which processes voltage signals of *measuring impedances* Z_{m1} and Z_{m2}, which are connected via the HV coupling capacitors C_{v1} and C_{v2} to the measured cable, which is connected to the test AC voltage. The advantage of this method is the elimination of interference, which would otherwise make it impossible to measure low charges of partial discharges in cables. In

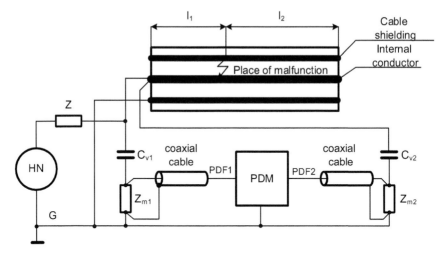

Figure 11.12   Differential galvanic method for measuring partial discharges of HV cables.

addition, the use of a two-channel partial discharge meter makes it possible to determine the fault locations in cables by the *refractometric method*.

Before the actual measurement of partial discharges, the cable is connected to a *charge calibrator*, which generates current pulses with an adjustable charge. These pulses propagate through the cable at speeds

$$v \approx \frac{1}{\sqrt{L \cdot C}} \left[\mathrm{ms}^{-1} \right], \tag{11.26}$$

which is determined by its inductance and capacitance per unit length.

If the cable insulation is faultless, the time delay of the reflected pulse of the open- or short-circuited other end of the cable is proportional to twice the cable length $l = l_1 + l_2$

$$\Delta T = \frac{2 \cdot l}{v}. \tag{11.27}$$

For example, time difference between generated and reflected pulse of faultless MV cable with specific inductance $L = 0.59$ mH/km and capacitance $C = 0.2$ μF/km, having propagation speed $v = 1.34 \cdot 10^8$ [ms^{-1}], is at its length $l = 1$ km $\Delta T = 7.4$ μs.

Figure 11.13 shows the time courses of current pulses in a cable with an open output, including its terminals.

If the cable insulation is damaged or the cable is short-circuited or broken, then the current pulse bounces off this damage point and its time delay is reduced to

$$\Delta T_1 = \frac{2 \cdot l_1}{v}. \tag{11.28}$$

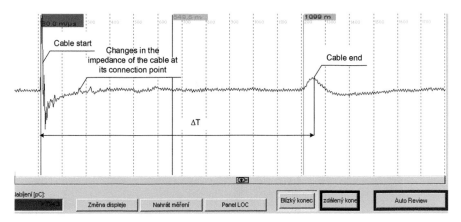

Figure 11.13 Time course of current pulses in a cable with open output.

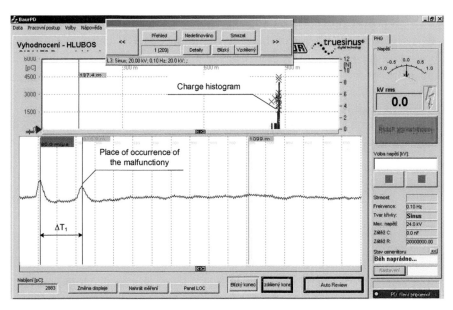

Figure 11.14 Determining the location of a cable failure, including the corresponding charge histogram.

Figure 11.14 shows the identification of the place of cable failure, including the corresponding charge histogram.

The measurement of partial discharges of HV cables is also related to the measurement of these discharges at their terminals.

For longer cable lengths, HV pulse generators with nominal voltages up to several tens of kilovolts are used to generate test pulses.

Figure 11.15   Breakdown of 110 kV VHV cable terminal due to partial discharges.

Cold or heat shrinkable terminals are used to connect the MV cables. Based on practical experience with the diagnosis of HV cables, it turns out that a significantly higher probability of failures in HV cable distribution does not occur in the cables themselves, but in their connection. Therefore, diagnostics of the condition of their terminals, e.g., by *ultrasonic, capacitive*, or *optical methods* of measuring *partial discharges*, is essential to increase the reliability of HV cable routes.

Figure 11.15 shows the breakthrough of the 110-kV HV cable terminal due to the action of partial discharges.

References

[1] Hauschild, W., Lemke, E. High Voltage Test and Measuring Techniques. Springer Verlag 2018. ISBN: 978-3-319-97459-0.
[2] Mentlík, V. *Diagnostics of electrical equipment.* Prague: BEN, 2008. ISBN 978-80-7300-232-9.
[3] Lucas, J. R.: High-Voltage-Cables. www.academia.edu/38036425/High_Voltage_Engineering.
[4] Vinter, P.: *Measurement of loss factor of power cables.*Elektro, 2001, č. 4.
[5] Shi, Q.,Troeltzsch, U.,Kanoun, O.:. *Detection and localization of cable faults by time and frequency domain measurements.* 2010 7th International Multi- Conference on Systems, Signals and Devices. IEEE, 2010, č. 1, http://ieeexplore.ieee.org/lpdocs/epic03/wrapper.htm?arnumber=5585506
[6] Baur: *Prüf- und Messtechnik GmbH. Cable Fault Location in LV, MV and HV Underground Cable Networks Practical Experience.* http://www.allbookez.com/pdf/276092/

[7] HV Technologies, Inc. *Cable Fault Location Measuring.* http://www.hvtechnologies.com/HVSolutions/CableFaultLocationSystems/CableFault LocationMeasuringMethods/tabid/328/Default.aspx#Top

[8] FCC Public s. r. o. *Diagnostic a testování silových kabelů.* http://www.odbornecasopisy.cz/index.php?id_document=26382

[9] EN 60 270 *High voltage testing technique - Partial discharge measurement.* 2001.

[10] FCC The Canadian Copper and Brass Development Association. http://coppercanada.ca/publications/pub23e/23e-section3.htm

[11] Industrija kablova - Jagodina (FKS). http://www.fks.co.rs/fkse/maticna/energet/upet/index110.htm

[12] Allkabel. http://www.allkabel.cz/high-voltage-cables-3630-kv-n2xsf2y-12-20-kv/

[13] Baur: Prüf- und Messtechnik GmbH. http://www.baur.at/fileadmin/ASSETS/brochures/CableFaultLocation/Kabelfehlerortu ng_de-de.pdf

[14] Kasztenny, B., Voloh, J., Jones, Ch. *Detection of incipient faults in underground medium voltage cables.* 2009 Power Systems Conference, IEEE, 2009, http://ieeexplore.ieee.org/lpdocs/epic03/wrapper.htm?arnumber=5262402

[15] Miyashita, Y., Y. Makishi a H. Kato. *Mechanism of water tree generation and propagation in XLPE.* http://ieeexplore.ieee.org/lpdocs/epic03/wrapper.htm?arnumber=172140

[16] Wen Shu, Jun Guo a S. A. Boggs. *Water treeing in low voltage cables.* IEEE Electrical Insulation Magazine, 2013, vol. 29, issue 2, s. 63-68. http://ieeexplore.ieee.org/lpdocs/epic03/wrapper.htm?arnumber=6457600

12

Diagnostics of Insulators, Surge Arresters, and Circuit Breakers

High-voltage insulators are designed for the insulation of HV overhead lines and for their mechanical fastening in substations and masts. HV arresters are used for pre-voltage protection of electrical equipment, caused, for example, by a lightning strike in an outdoor HV line. *Power switches* are used to switch on and off parts of MV circuits in MV substations and overhead lines.

12.1 HV Insulators

We distinguish between ceramic, glass, and composite insulators according to the material used. Depending on the type, we distinguish between suspension, anchor, and support insulators.

Figure 12.1 shows individual types of HV insulators.

Ceramic insulators are made of electrical porcelain, which is a mixture of quartz, feldspar, and kaolin. After their shaping and firing, the insulators are provided with a glass glaze. This prevents their moisture and increases their surface insulation resistance. The advantage of ceramic insulators is their low price and high dry electrical strength. Their disadvantage is that water droplets adhere to them, which reduces their surface resistance in the rain.

Glass insulators are made of quartz, sodium carbonate, and limestone. For voltage levels above 22 kV, glass insulators are made of boron glass. Due to ceramic insulators, glass insulators have higher electrical strength and are also cheaper.

Composite insulators consist of a metal armature on which a supporting insulating rod with a silicone shell and with silicone rubber roofs is threaded. Their advantages are high *hydrophobicity*, i.e., the ability to repel water, and low weight.

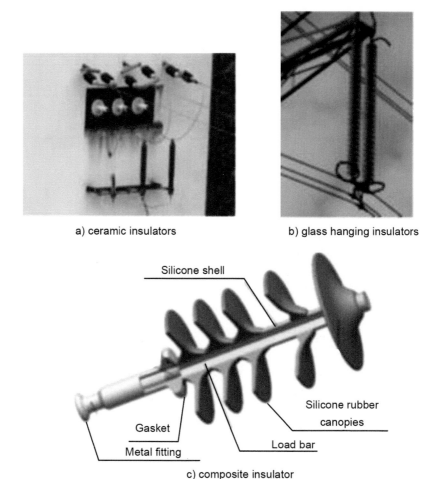

a) ceramic insulators   b) glass hanging insulators

c) composite insulator

Figure 12.1   Types of insulators.

12.1.1 Parameters of HV insulators

The basic electrical parameters of the insulators include their maximum operating voltage, jump and surface distance, distance between the insulator roofs, shape factor, diameter of the support rod, diameter and distribution of the roof, and the wall thickness of the insulator.

The *operating voltage* of the insulators is determined by the RMS value of the voltage that the insulator is able to insulate from other supply or ground conductors. The insulators are manufactured in standardized voltage levels of 3, 6, 10, 22, 35, 110, and 220 kV.

The *isolation leakage L* is defined as the shortest distance between the live parts of the insulator and its grounded portion.

The *distance creepage* of the insulator is the shortest distance on the surface of the insulator between its two conductive parts, respectively between live and ground conductors.

The *form factor* determines the ratio between the distance between the insulator's roofs and the depth of its roof. This is true for this factor

$$FF = \int_0^L \frac{dl}{2\pi \cdot R(l)}, \tag{12.1}$$

where L is the surface distance and R is the radius of the insulator, measured from the center of the rod to the outermost edge of the roof.

The *form factor* can be used to compare the properties of insulators with the same length, diameter, and surface distance but with different profiles, i.e., the shape and spacing of roofs. However, since we usually do not know the exact functions that would accurately describe the curvature and shape of individual roofs, this shape factor is usually only an approximate parameter of the insulators.

The *shaft diameter* d_{sh} rod corresponds to the diameter of the narrowest point on the entire insulator, which is usually the support rod. This parameter determines the overall ability to withstand the mechanical load of the insulator. For composite insulators, this parameter can also be expressed by the thickness of the glass fiber bundle forming the support rod.

The *shed diameter* D_s is the diameter of the widest part of the insulator. This parameter is important for assessing the voltage resistance of insulators during their contamination.

Insulator wall thickness t is given for hollow insulators, where there is a risk of electrical breakdown of their walls by internal voltage jump due to inhomogeneity of the insulator material.

Figure 12.2 shows a cylindrical HV insulator with its parameters.

12.1.2 Voltage tests of insulators

Standard tests of HV insulators are *voltage tests with AC voltage* in dry and wet conditions, and tests with *atmospheric* and *switching pulses* in dry and rainy conditions. In all these tests, the test conditions are simulated as faithfully as possible to match the operating conditions of the insulators. Therefore, the entire insulation kits and insulator hinges, including their mounting on parts of electric poles, are implemented in the insulator test rooms.

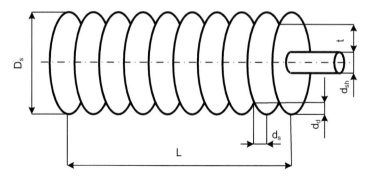

Figure 12.2   HV cylindrical insulator.

12.1.2.1 Dry stress tests

These tests are performed by *atmospheric* or *switched pulse*, and the test voltage is corrected to atmospheric conditions. The reference atmospheric conditions are considered to be temperature $\vartheta_0 = 20°C$, tlak $p_0 = 1013$ hPa, and absolute humidity of 11 g/m³.

The magnitude of *the test voltage* with respect to atmospheric conditions is determined by the following equation:

$$U = k_t \cdot U_0 = k_1 \cdot k_2, \tag{12.2}$$

where k_t is the atmospheric correction factor, k_1 is the air density correction factor, and k_2 is the humidity correction factor.

The air *density correction factor* depends on its relative density

$$k_1 = \delta^m, \tag{12.3}$$

where m is the exponent depending on the type of discharge and

$$\delta = \frac{p}{p_0} \cdot \frac{273 + \vartheta_0}{273 + \vartheta}, \tag{12.4}$$

where ϑ_0 and p_0 are the reference temperature and air pressure, and ϑ and p are its current temperature and pressure.

The *humidity correction factor* is a defined equation

$$k_2 = k^w, \tag{12.5}$$

where w is the exponent, depending on the type of discharges.

Applies to *DC test voltage*

$$k_{2DC} = 1 + 0.012\left(\frac{h}{\delta} - 11\right). \tag{12.6}$$

Applies to *AC test voltage*

$$k_{2\,AC} = 1 + 0.010\left(\frac{h}{\delta} - 11\right). \tag{12.7}$$

The exponents m and w are determined by the parameter g, which depends on the burst voltage U_{50} at the atmospheric conditions under which the voltage tests are performed.

$$g = \frac{U_{50}}{500 \cdot L \cdot \delta \cdot k}, \tag{12.8}$$

where L is the minimum discharge path in m.

In Figure 12.3, the dependences of the parameters w and m on the parameter g are given.

For example, when recalculating the AC test voltage $U = 123$ kV, measured at a temperature $\vartheta = 32\ °C$, atmospheric pressure $p = 998$ hPa and humidity $R = 69.8\%$, the standard value of the test voltage $U_0 = 162.2$ kV is based.

12.1.2.2 Wet stress tests

These tests are performed in exactly the same way as the dry tests, with the difference that artificial rain falls on the surface of the insulators. It is recommended to perform the measurement for all types of test voltage. The tested objects are sprayed with water with a defined resistivity and a defined temperature. It is required that no mist or splinters form during spraying and that the drops correspond as much as possible to the actual rain.

In addition, the minimum distance of the test object from the surrounding objects and the structure must be observed, which should be at least 1.5 times the length of the shortest possible discharge path, in order to avoid shocks on the surrounding objects, which would distort the test results.

The voltage distribution along the test object and the electric field of the supply electrodes, which must be independent of external influences, must also be taken into account. Before the start of the measurement, the object should be irrigated with treated water for at least 15 minutes, or with untreated water for 15 minutes and then with treated water for another 2 minutes. During the tests, the same conditions as for raining must be observed and the rain must not be interrupted in any case.

Artificial rain should resemble real rain as much as possible. The temperature of the artificial rain should roughly correspond to the ambient temperature, and the permissible deviation of the water temperature is ±15°C. For spraying the tested insulators, the precipitation equivalent must be

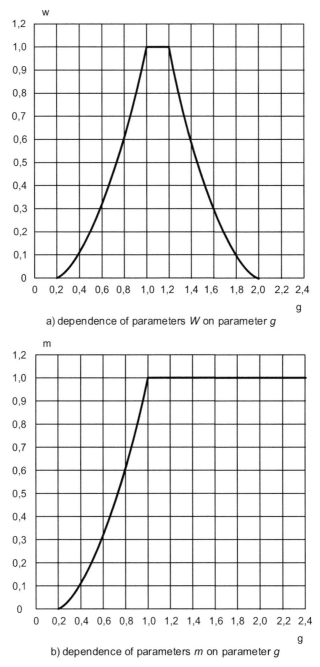

a) dependence of parameters *W* on parameter *g*

b) dependence of parameters *m* on parameter *g*

Figure 12.3   Dependence of parameters *w,m* on parameter *g*.

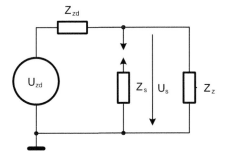

Figure 12.4 Principle of surge arrester operation.

1–2 mm/min, i.e., 60–120 mm/h with a maximum deviation of ±0.5 mm/min, which corresponds to very heavy rain. The horizontal and vertical components of the rain must be approximately the same. The intensity of the rain is controlled by means of a collecting vessel with separate chambers.

12.2 Preload Arresters

Surge arresters are devices used to protect high-voltage electrical. They are always connected in parallel to the protected device (Figure 12.4).

If the HV line is not under a voltage higher than its maximum operating level, then the impedance of the voltage arrester Z_s is significantly higher than the internal impedance of the source Z_{zd} and the impedance of the load Z_z. If a voltage higher than its rated voltage is induced in the line, then the internal resistance of the voltage arrester drops to a value that is significantly less than the load impedance. In this case, the current flowing through the surge arrester has a value

$$I_s = \frac{U_{zd}}{Z_s + \dfrac{Z_{zd} \cdot Z_z}{Z_{zd} + Z_z}}. \qquad (12.9)$$

The following applies to the voltage on the arrester:

$$U_s = \frac{Z_s \cdot (Z_{zd} + Z_z)}{Z_s \cdot (Z_{zd} + Z_z) + Z_{zd} + Z_z} U_{zd}. \qquad (12.10)$$

12.2.1 Types of surge arresters

From a design point of view, we recognize air spark gaps, blow-off lightning arresters, valve lightning arresters, and voltage limiters.

12.2.1.1 Air spark gap

The *air spark gap*, also referred to as the protective spark gap, is in the simplest embodiment formed by two bent conductors, which are spaced apart by a *landing distance d* (Figure 12.5).

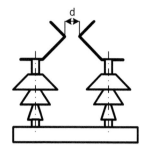

Figure 12.5   Air spark gap.

If the voltage at the spark gap terminals does not exceed its ignition voltage, then there is a very high impedance between its electrodes. If the voltage at the terminals of the spark gap exceeds the *ignition voltage*, an arc discharge is created between its electrodes, which causes the impedance between the poles of the *spark gap to jump* to a low value close to zero and the voltage at the spark gap drops to several tens of volts.

Rebound distances $d = 12$ cm are used for the nominal network voltage of 22 kV, and plate distances $d = 15$ cm are used for the voltage of 35 kV.

The advantage of these spark gaps is simplicity in principle. Their disadvantage is the steep drop in voltage after the ignition of the discharge, which puts unreasonable strain on the insulation system of electrical machines and thus shortens their service life. Another disadvantage of these spark gaps is the large dispersion of ignition voltages (±15%).

12.2.1.2 Blow arrester

The *blowing lightning arrester* consists of an air spark gap, which has the function of a disconnector and a closed internal spark gap for extinguishing the electric arc (Figure 12.6).

The internal spark gap is enclosed in a tube of gaseous substance. Fiber used to be used as a gaseous substance; today, *Umaplex organic glass (polymethyl methacrylate)* is used. During normal operation, the closed spark gap is de-energized, but when an overvoltage occurs, the air in the closed spark gap ignites with the help of an air spark gap. The heat of the arc generates heat inside the tube, which releases gases. These gases open the lid when their certain pressure is exceeded. The flowing gases cool the arc on the spark gap; so the arc of the open spark gap goes out.

These so-called *Torok tubes* are used in HV distribution networks up to a nominal voltage of 22 kV. Blow arresters have less ignition delay than spark gaps but larger than valve arresters. The advantage of these lightning arresters is the low purchase price and the blowing of the arc away from the conductors. This is also related to the disadvantage of lightning arresters because the arc can reach several meters during blowing. The disadvantage of these lightning arresters is their limited service life of 20–30 discharges. Therefore, they are currently being replaced by valve lightning arresters.

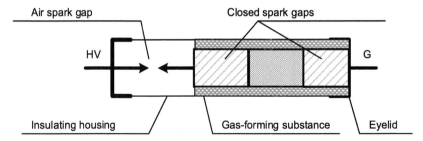

Figure 12.6 Blow plugs.

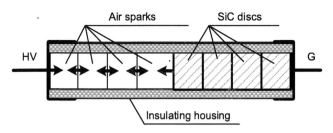

Figure 12.7 Valve lightning arrester.

12.2.1.3 Valve lightning arrester

The *valve lightning arrester* consists of several air sparks connected in series and a nonlinear series resistor (Figure 12.7).

The *landing distances* of the series-connected air *spark gaps are* in the order of tenths of millimeter, so that the voltage jumps take place as quickly as possible. The advantage of series arrangement of spark gaps is also their more efficient cooling and deionization of arcs, which guarantees good extinguishing properties. Nonlinear series resistors are composed of cylindrical disks with a metallized surface (Ag, Al, and Cu-Zn). The main material is silicon carbide (SiC). Its V–A characteristic is strongly nonlinear.

As a large discharge current passes through the lightning arrester, the resistance of the disks becomes so low that it does not allow the voltage at the lightning arrester to exceed the ignition voltage. When an overvoltage occurs on the protected equipment, the spark gap ignites and current flows into the ground through the SiC disks. As the overvoltage increases, the resistance of the disks decreases and the current flowing to the ground increases. However, the magnitude of the resistance must be large enough so that the current flowing into the ground is small enough and can interrupt in the spark gap when the current passes through zero. The resistance discs are de-energized after a current interruption in the spark gap (no current flows into the ground).

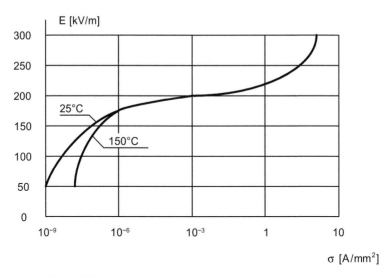

Figure 12.8   VA characteristics of resistance disc based on ZnO.

Valve lightning arresters are used in one-part MV networks, in several-part HV networks connected in series. Lightning arresters with a nominal voltage from 110 kV are supplemented by a protective ring. This ensures an even distribution of the operating voltage to the parts of the lightning arrester, thus preventing the surge voltage from being so great.

The advantage of the valve lightning arrester is that there is no steep voltage drop during the jump to the spark gap, which endangers the insulation of electrical machines. Their disadvantage is that if the current in the spark gaps does not go out when the current passes through zero for the first time and the SiC blocks do not withstand the absorption of thermal energy, then the lightning arrester explodes. Another disadvantage is its complex internal construction.

12.2.1.4 Non-sparking preload limiters

Non-sparking overvoltage limiters are formed by resistance disks made of *zinc oxide* (ZnO) with an admixture of *bismuth dioxide* (Bi_2O_3), which have a significant nonlinear course of their V–A characteristics (Figure 12.8).

In the middle part of this characteristic, its nonlinearity can be expressed by the following equation:

$$I = k \cdot U^{\alpha}, \tag{12.11}$$

where k is the constant of the voltage level for which the arrester is intended, and α is the exponent, acquiring values of 20–51 for ZnO (for SiC blocks in valve lightning arresters, it is $\alpha = 2$–6).

Table 12.1 Breakdown of bias limiters according to absorbed energy.

Energy Class	Rated voltage [kV]	Rated discharge current [kA]	Current overload capability [kA]	Absorbed energy [kJ/kV]
1	3–51	10	20	2.8
2	6–120	10	40	4.5
3	6–360	10	50–63	6.7
4	6–240	20	50–63	9.2

For example, at $\alpha = 51$, increasing the voltage by 20% increases the current by more than four orders of magnitude, i.e., from 100 mA to 1 kA. For currents higher than 1 kA, the nonlinearity of the V–A characteristic decreases.

Preload limiters are divided according to the ability to absorb energy generated by their activity into four energy classes, according to Table 12.1.

The advantage of ZnO preload limiters is their very fast response to preload and maintenance-free operation. Their relatively simple construction guarantees high mechanical resistance, and thus resistance to vibration and shock.

Figure 12.9 shows typical HV spark-free preload limiters.

12.2.2 Parameters of preload limiters

The basic parameters of the bias limiters include the continuous operating voltage, the rated voltage, the rated discharge current, and the residual voltage.

The *continuous operating voltage* U_c is the highest voltage that can be permanently connected to the limiter with respect to its thermal stress. The *rated voltage* U_r is the highest voltage so that the limiter works correctly even in conditions of temporary overvoltages. The equation $U_r = 0.8 \cdot U_c$ applies between these voltages.

The *rated discharge current* is the current that the limiter is able to conduct. The residual voltage is the voltage corresponding to the rated discharge current. Thus, current pulses can be 1/20 and 8/20 μs and current pulses with a rise time in the range of 30–100 μs and with half-life times of 60–200 μs.

12.2.3 Voltage tests of voltage arresters

The standard EN 60 099 dry the ignition voltage tests, 50-Hz AC voltage tests, residual voltage measurements, current pulse endurance test, long

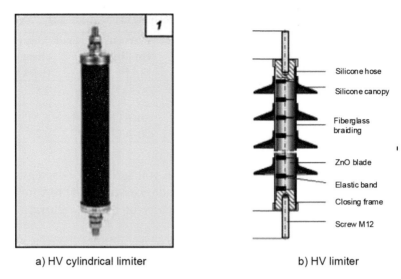

a) HV cylindrical limiter   b) HV limiter

Figure 12.9   Non-sparking preload limiters.

current pulse endurance test, operational functionality test, and tests using partial discharges. Other tests of voltage arresters are their mechanical tests, tests of the influence of the environment, and tests of radio interference.

12.2.3.1 Ignition voltage tests

This test is mandatory for all types of surge arresters. It is performed in dry and rainy conditions with an alternating voltage having a frequency of 50 Hz, while the test voltage increases until the value when the arrester responds. The power supply must be switched off within 0.5 s of the arrester's response. Furthermore, the arrester is tested with an *atmospheric pulse* of 1.2/50 μs with a peak voltage value equal to 2.5 times the nominal voltage of the arrester. During this test, five pulses of positive and negative values are applied, and the surge arrester must respond.

When measuring the voltage characteristic of the ignition voltage, atmospheric pulses with increasing peak values are gradually applied, and the times at which the surge arrester reacts are read. *Switching pulses* are used to measure voltage characteristics for arresters from 100 kV and 10 kA.

12.2.3.2 Residual voltage measurement

It is also performed at atmospheric pulses 8 μs/20 μs, while at least four pulses with peak current values of 0.25, 0.5, 1, and 2 times the rated current

Table 12.2   High current pulse peaks.

Lightning arrester class (rated discharge current) [kA]	Peak value of high current pulse (nominal discharge current) [kA]
10	100
5	65
2.5	25
1.5	10

of the arrester are applied to the arrester. There must be such time delays between its individual discharges that the arrester cools down to ambient temperature. The result of the measurement is the dependence of the residual voltage on the discharge current.

12.2.3.3 Current pulse endurance test

This test is performed on three samples of arresters that are assembled for operational use. There must also be such a delay between their individual discharges, as when measuring the residual voltage, so that the arresters cool down to ambient temperature. Dry ignition measurements shall be measured before and after this test. We distinguish between endurance pulse current tests with high current pulse and long current pulse tests.

The high *current pulse endurance test* is performed with two 4/10 μs current pulses with the peak values given in Table 12.2.

At each of these pulses, the voltage and current must be measured, and these values should not differ significantly for the lightning arresters tested.

The *long-current impulse endurance test* is performed with 20 discharges, which are divided into five groups after five discharges. The time delay between discharges in each group must be 50–60 s and between groups must be 25–30 minutes.

12.2.3.4 Operational functionality tests

It is performed on samples on which endurance tests have not been performed. As in the previous case, 20 pulses are applied in four groups of five pulses, and the time delay between discharges in each group must be 50–60 s and between groups must be 25–30 minutes.

The *rated AC test voltage* must be applied for at least one period, before the application of the test pulses and at least 10 s after their application. In each case, the subsequent pulse must be interrupted at the latest by the end of the half-period of the AC voltage following the half-period in which the current pulse was applied. Subsequent current must never be re-generated in the following half-periods of the AC voltage.

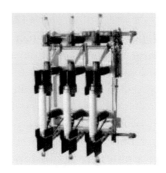

a) circuit breakers with FS₆   b) compression switch-disconnectors

Figure 12.10   Power switch and switch-disconnector.

12.3 Power Switches and Disconnectors

Power switches are designed for switching on and off parts of MV and HV circuits, through which not only rated but also short-circuit currents can flow. *Switch-disconnectors* are used for current-free disconnection of these circuits, and, therefore, their contacts are not designed for rated currents.

Figure 12.10 shows a typical SF_6 gas-filled circuit breaker and a compression switch for a nominal voltage of 22 kV.

The main parts of the circuit breakers are switching contacts, extinguishing chambers, and a trigger, which automatically disconnects this circuit in the event of a short circuit. For large MV circuit breakers, sulfur fluoride (SF_6) or compressed nitrogen is used to suppress arcing between the contacts. Circuit breakers must meet strict requirements for reliability, resistance to dirt, and constant load.

12.3.1 Diagnostic methods for circuit breakers

Voltage tests of the electrical strength of the power circuit, measurement of the resistance of the power circuits, measurement of partial discharges at their bushings, warming tests, on and off tests, and short-term and dynamic current tests are used to test the properties of circuit breakers.

In accordance with the EN 60 060 standard, voltage tests of electrical strength are performed on both sides of an open switch by an atmospheric or switching pulse.

The *resistance of the power circuit* is measured by the Ohm method before any temperature tests with a direct current in the range up to 50 A, while the voltage drop across the closed switches is measured. This resistance is in the order of tens of megaohms. A similar test is performed for auxiliary contact circuits, using direct current up to 10 mA and voltage up to 6 V. The measured resistances of the auxiliary contacts should not exceed 50 Ω.

Switch-on and *switch-off tests*, which belong to functional tests, determine the speed and quality of switching on and off of power switches when switching resistive, capacitive, and inductive loads, including repeated switching.

Warming tests are performed in closed rooms with the exclusion of air flow at rated flow and the temperature is expected to stabilize with a time change of 1°C/hour. During this test, the air temperature should not change by more than 10°C.

Short-term and *dynamic withstand current tests* determine the resistance of power switches to the effects of short-term withstand current and rated dynamic withstand current. The rated short-time withstand current is the current that the switch can withstand for the required short time. The rated dynamic withstand current is the maximum value of the current amplitude that the switch can withstand without damage. The rated short-circuit time is the time for which the closed switch can transmit the rated short-time withstanding current for 1 s without damage.

These tests are performed at a nominal frequency of 50 Hz. Before each test, a no-load measurement shall be made to measure the resistance of the main circuit (except for earthing switches). The test can be single-phase or three-phase (a single-phase test tests two pairs of adjacent poles in series).

References

[1] Hauschild, W., Lemke, E. High Voltage Test and Measuring Techniques. Springer Verlag 2018. ISBN: 978-3-319-97459-0.

[2] Farzaneh, M. - Chisholm W. *Insulators for icing and polluted environments*. Hoboken, NJ: J. Wiley, c2009, xxvi, 680 p. ISBN 0470282347

[3] Kuffel, E, W Zaengl a J Kuffel. *High voltage engineering: fundamentals*. 2nd ed. Boston: Butterworth-Heinemann, 2000. xiii, 539 p.

[4] *Compozite isolators VHV Maclean Power Systems*.

[5] EN 60060-1. High-voltage test techniques. General definitions and test requirements, 2010.

[6] EN 60060 – 2. High-voltage test techniques. *Masuring system*. 2012.

[7] EN 60383-2. *Insulators for outdoor lines with a nominal voltage above 1000 V. Part 2: Insulator chains and insulator hinges for AC systems. Definitions, test methods and acceptance criteria*. 1996.

[8] Kotlan, V. *Impact Phenomena on Transmission Lines*, Dissertation, University in Pilsen 2008.

[9] Farzaneh, M. - Chisholm W. *Insulators for icing and polluted environments.* Hoboken, NJ: J. Wiley, c2009, xxvi, 680 p. ISBN: 0470282347.

[10] Kuffel, E, W Zaengl a J Kuffel. *High voltage engineering: fundamentals.* 2nd ed. Boston: Butterworth-Heinemann, 2000.

[11] Catalog DRIBO, Brno: Compozite isolators *VHV Maclean Power Systems.*

[12] Insulators 101. http://ewh.ieee.org/soc/pes/iwg/Publications/ Insulators 101PanelFinalA.ppt.

List of symbols

AC	Alternating Current
ADC	Analog Digital Convertor
AIT	Atmospheric Impulse Test
AM	Amplitude Modulation
CCD	Charge Coupled Devices
CLVT	Constant Level Voltage Test
DA	Dielectric Absorption
DAC	Digital Analog Convertor
DC	Direct Current
DFT	Discrete Fourier Transform
DGA	Dissolved Gas Analysis
DUT	Devices Under Test
FDS	Frequency Domain Spectroscopy
FRA	Frequency Response Analysis
GST	Grounded Specimen Test
GSTQ	Grounded Specimen Test Quarding
HV	High Voltage
IFRA	Impulse Frequency Response Analyzing
LTVT	Long Term Voltage Test
LV	Low Voltage
MEMS	Micro Electro Mechanical Systems
MSS	Mean Square Sum
PCB	Polychlorbiphenyl
PDM	Partial Discharge Meter
PE	Polyethylene
PI	Polarization Index
Sps	Sample per Seconds
PVC	Polyvinylchloride
RMS	Root Mean Square
SFRA	Sweep Frequency Response Analyzing
SIT	Switching Impulse Test
STV	Step Voltage Test

SV	Small Voltage
UDVT	Up–Down Voltage Test
UHF	Ultra-High Frequency
UHV	Ultra-High Voltage
UST	Unground Specimen Test
VHV	Very High Voltage
VLF	Very Low Frequency Test
WVT	Withstand Voltage Test

List of Constants and Quantities

Mathematical Constants

Euler's number	2.718
Ludolf's number	3.1415926535

Physical Constants

Avogadro's constant	$6.0221 \cdot 10^{23}$ [mol^{-1}]
Planck's constant	$6.626472 \cdot 10^{-34}$ [J.s]
Boltzmann's constant	$1.38 \cdot 10^{-23}$ [J·K^{-1}]
The speed of light	$3 \cdot 10^{8}$ [m·s^{-1}]
Electron charge	$1.60234 \cdot 10^{-19}$ [C]
Electron mass	$9.1 \cdot 10^{-31}$ [kg]
Earth acceleration	9.81 [m·s^{-2}]
Vacuum permittivity	$8.854 \cdot 10^{-12}$ [F·m^{-1}]
Vacuum permeability	$1.2563 \cdot 10^{-6}$ [H·m^{-1}]

Mechanical Quantities

Quantities	mark	dimension
Mass	m	[kg]
Time	t	[s]
Deflection	x	[m]
Speed	v	[ms^{-1}]
Acceleration	a	[ms^{-2}]
Shock	s	[ms^{-3}]
Force	F	[N] = [kg·ms^{-2}]
Moment	M	[N·m]
Viscosity	n	[m^{2}s^{-1}]
Rigidity	k	[N·m^{-1}]
Damping	b	[Pa] = [N·m^{-2}]
Pressure	p	[Pa] = [N·m^{-2}]
Modulus of elasticity	G	[N·m^{-2}]

Angle	ϕ	[rad]
Frequency	f	[Hz]
Angular frequency	ω	$[s^{-1}]$
Period	T	[s]
Energy	E	$[J] = [W \cdot s]$

Thermal Quantities

Quantities	**Mark**	**Dimension**
Absolute temperature	θ	[K]
Celsius temperature	ϑ	[°C]

Electrical Quantities

Quantities	**mark**	**dimension**
Current	I	[A]
Voltage	U	[V]
Active power	P	[W]
Reactive power	Q	[VAR]
Apparent performance	S	[VA]
Power factor	$\cos\phi$	–
Charge	Q	$[C] = [A \cdot s]$
Current density	J	$[A \cdot m^{-2}]$
Electrical induction	D	$[C \cdot m^{-2}]$
El. intensity field	E	$[V \cdot m^{-1}]$
Magn. inductive flow	Φ	$[Wb] = [V \cdot s]$
Intensity magn. field	H	$[A \cdot m^{-1}]$
Magnetic induction	B	$[T] = [V \cdot s \cdot m^{-2}]$
Resistance	R	$[W] = [V \cdot A^{-1}]$
Impedance	Z	$[W] = [V \cdot A^{-1}]$
Admittance	Y	$[S] = [A \cdot V^{-1}]$
Capacity	C	$[F] = [A \cdot s \cdot V^{-1}]$
Inductance	L	$[H] = [V \cdot s \cdot A^{-1}]$
Resistivity	ρ	$[W \cdot m]$
Conductivity	γ	$[S \cdot m^{-1}]$
Permittivity	ε	$[F \cdot m^{-1}]$
Permeability	μ	$[H \cdot m^{-1}]$

List of Standards

Standard number	Content of the standard	Valid from
EN 60076	Power transformers. In general	2012
EN 60052	Voltage measurement using standardized air spark gaps	2003
IEC 60061-1	High voltage test technique. Part 1: General definitions and test requirements	2010
IEC 60060-2	High voltage test technique. Part 2: Measuring systems	2011
EN 60099	Surge arresters	2018
EN 60243-3	Electrical strength of insulating materials. Test methods. Part 3. Additional requirements for impulse tests 1.2/50 ms	2002
EN 60270	High voltage test technique. Measurement of partial discharges	2002
EN 60383	Insulators for overhead lines with a rated voltage above 1 kV. Part 2: Insulator chains and insulator hinges for AC systems. Definitions, test methods, and acceptance criteria	1996
EN 61109	Insulators for overhead lines. Composite suspension and anchor insulators for AC systems with a rated voltage greater than 1 kV – definitions, test methods, and acceptance criteria	2009
EN 61180	High voltage test technique for low voltage equipment. Definitions, test requirements, test procedures, and test equipment	2017
EN 61198	Mineral insulating oils. Methods for determination of related compounds	1997
EN 61952	Insulators for overhead lines. Composite support insulators with a rated AC voltage exceeding 1 kV	2003
EN 62217	Polymeric insulators for outdoor and indoor use with a rated voltage > 1 kV – general definitions, test methods, and acceptance criteria	2006
EN 62223	Insulators. Terms and definitions	2010
EN ISO 12937	Petroleum products. Determination of water. Coulometric titration method according to K. Fischer	2003

Standard number	Content of the standard	Valid from
EN ISO 3104	Petroleum products. Transparent and opaque liquids. Determination of kinematic viscosity and calculation of dynamic viscosity	1998
EN ISO 62021-1	Insulating liquids. Determination of acidity by potentiometric method	2004
IEC 250	Recommended procedures for determining the permittivity and loss factor of electrical insulating materials at industrial, acoustic, and radio frequencies, including meter lengths	1998
IEC 346210	Basic aspects of functional evaluation of electrical insulation systems. Mechanical aging and diagnostic procedures	1996
IEC 354	Instructions for loading power transformers	1997
IEC 3831	Insulators for overhead lines with a rated voltage above 1 kV. Part 1. Ceramic or glass insulators for AC systems. Definitions, test methods, and acceptance criteria	1996
IEC 493-1	Guideline for statistical analysis of aging test data. Part 1. Methods based on the mean values of normally distributed test results	1996
IEC 554-1+A1	Specification for cellulosic papers for electrical purposes. Part 1. Definitions and general requirements	1983
IEC 60034-18-1	Rotating electrical machines. Part 18. Functional evaluation of insulation systems	1996
IEC 60085	Electrical insulation. Thermal evaluation and marking	2008
IEC 60216-6	Electrical insulating materials. Heat resistance properties. Part 6. Determination of thermal resistance indices of insulating material by fixed times method	2004
IEC 60505	Evaluation and classification of electrical insulation systems	2005
IEC 62068-1	Electrical insulation systems. Electrical stresses caused by repeated impulses	2004
IEC 62068-1	Electrical insulation systems. Electrical stresses caused by repeated impulses	2004
IEC 6216-1	Electrical insulating materials. Heat resistance properties. Part 1: Aging process and evaluation of the test result	2002
IEC 836	Specification for silicone fluids for electrical purposes	1997
EN 61063	Measurement of airborne noise emitted by steam turbines and their engines	1991

Note:

EN is a designation of harmonized standards by the *European Committee for Standardization.*

The IEC (*International Electrotechnical Commission*) is a global organization founded in 1906 that develops and publishes international standards for electrical engineering, electronics, communications, and related industries.

ISO (*International Organization for Standardization*) is an international organization for standardization.

Index

About the Author

In 1971 Josef Vedral graduated from the Czech Technical University in Prague, Faculty of Electrical Engineering in the field of measuring technology. In 1976 he completed his doctoral studies at the Department of Measurement of the Czech Technical University in Prague in the field of measuring instrumentation. In recent years, he has been engaged in condition monitoring of HV electric machines. Between 2012 and 2020 he was the head of two projects of the Technology Agency of the Czech Republic projects "Intelligent measuring diagnostic system for estimating the operating condition of high-voltage electric rotary and non-rotating machines" and the related project "Compact diagnostic system for monitoring the condition of high-voltage electric machines using DC and low-frequency alternating test voltage". He is the author of 82 scientific articles in international journals, 12 Czech patents, 12 university textbooks in Czech and 2 textbooks in English.